PSYCHIC
INVESTIGATORS

SCIENCE & CULTURE IN THE NINETEENTH CENTURY

Bernard Lightman, Editor

PSYCHIC INVESTIGATORS

ANTHROPOLOGY, MODERN SPIRITUALISM, &
CREDIBLE WITNESSING
IN THE LATE VICTORIAN AGE

EFRAM SERA-SHRIAR

UNIVERSITY OF PITTSBURGH PRESS

Published by the University of Pittsburgh Press, Pittsburgh, Pa., 15260
Copyright © 2022, University of Pittsburgh Press
All rights reserved
Manufactured in the United States of America
Printed on acid-free paper
10 9 8 7 6 5 4 3 2 1

Cataloging-in-Publication data is available from the Library of Congress

ISBN 13: 978-0-8229-4707-3
ISBN 10: 0-8229-4707-2

Cover design by Alex Wolfe

To Rupert Shriar, my beloved grandfather. May he rest in peace.

CONTENTS

CONTENTS

EPILOGUE

Legacies of Late Victorian Spirit Investigations

159

ACKNOWLEDGMENTS

Since embarking on this project in January 2016, I have received tremendous help and support from many friends and colleagues. This book would not have been possible without you. I want to begin by thanking my former colleagues at the Leeds Centre for Victorian Studies; particularly, Nathan Uglow, Rosemary Mitchell, Josh Poklad, Karen Sayer, Di Drummond, Jane de Gaye, and Kate Lister. My gratitude also extends to the Centre for History and Philosophy of Science at the University of Leeds. Graeme Gooday, Greg Radick, Jonathan Topham, Jamie Stark, Mike Finn, and Adrian Wilson were especially supportive. I have also had the pleasure of working with many other fine scholars across the academic world, who at various stages of the book's development provided useful guidance and information. The following group of people deserve special mention: Geoffrey Cantor, Christine Ferguson, Roger Luckhurst, Shane McCorristine, Elsa Richardson, Peter Lamont, Andreas Sommer, Emily-Jane Cohen, Bennett Zon, Richard Noakes, Maureen Meikle, Suzanne Owen, Rebecka Klette, Jamie Elwick, Iwan Morus, Jonathan Livingstone-Banks, Geoff Belknap, Chris Manias, Abigail Woods, Casper Andersen, and Ian "the Hammer" Hesketh. During my time at the Science Museum in London between January 2019 and November 2021, the Department for Research and Public History was wonderfully accommodating and encouraging. My thanks extend to them all, but Tim Boon deserves particular acknowledgment.

I also want to thank Abby Collier at the University of Pittsburgh Press for her commitment to and support of the project. I am equally grateful to the invaluable feedback I have received from my anonymous referees at various stages of the project's development. No person has done more for me over the past decade than Bernie Lightman. He has been a wonderful friend, and a tremendous mentor. Thank you for always being there when I needed you. Finally, my amazing wife, Nanna Kaalund, and our adorable (and precocious) son, Magnus, were tremendously patient throughout my writing of this book, and brought me much joy during the more stressful periods. You guys rock!

PSYCHIC
INVESTIGATORS

Introduction

When Victorian Anthropology Entered the Séance Room

On November 26, 1866, the ethnologist-turned-anthropologist Edward Burnett Tylor (1832–1917) wrote a letter to his friend the naturalist Alfred Russel Wallace (1823–1919), on the topic of modern spiritualism. Tylor had recently discovered that Wallace was a supporter of the spirit hypothesis—namely, the idea that the phenomena produced by mediums at séances are caused by disembodied spirits. Skeptical of the published accounts by eyewitnesses that he had been reading in books and periodicals professing to have observed authentic spirit manifestations, Tylor argued that he would put much more trust in the reports of a skilled scientific observer such as Wallace, if he were to publish the details of his experiments investigating spirits. Tylor wrote, "I believe you really know how to observe and describe, why do you not make a series of direct investigations yourself with all precautions against imagination and fraud? Your results would, I believe, be thought more of, whatever results might lead you to, than all the spiritualistic literature extant."[1]

This letter between Tylor and Wallace brings to the fore a key issue within nineteenth-century debates over the existence of spirits, and the evidence for their manifestation at séances: who should be trusted as a credible witness of spirit and psychic phenomena? It is my aim in this book to explore this question by examining anthropology's engagement with modern spiritualism during the late Victorian era. Rather than being a mere footnote to the discipline's history, spiritualism posed a genuine concern for nineteenth-century anthropological researchers, and required serious and rigorous investigation. If the spirit hypothesis were proven to be true, it would cut to the core of British anthropology's foundation.

Modern Spiritualism and Late Victorian Anthropology

Traditionally, scholars have attributed the rise and growth of modern spiritualism over the past two centuries to the so-called nineteenth-century crisis of faith.[2] As scientific and medical explanations of the natural world began to displace older religious ones, there was a rise in secular thinking. Naturalistic explanations for the origin of human life, in particular, led to a critical reassessment of orthodox religious ideas, including special creation, the age of the Earth, and the existence of heaven and hell. The emergence of biblical criticism was an equally powerful source of social subversion, allowing Victorians to question scriptural authority.[3] It was within this cultural climate that spiritualism's appeal grew, as many nineteenth-century figures appropriated it as a way of reconciling their fears about the afterlife and the loss of traditional religious faith. The discovery of a solar heat death in 1852 by the mathematician and physicist William Thomson, First Baron Kelvin (1824–1907), raised further intellectual and social anxieties. If the entire material world was to be destroyed by a solar implosion, what would remain of humanity's legacy? How would people be remembered if something physical no longer existed?[4] Modern spiritualism, with its professed commitment to nonmaterial governing laws akin to those in nature, offered a potential solution. If spirits were proven to be real, it would provide clear evidence in favor of an existence beyond death and the material world. These sorts of preoccupations firmly rooted spiritualism (and by extension psychical research) in epistemic debates about the authority and limits of both science and religion.[5]

Looking back on the history of the exposure of fraudsters in modern spiritualism, it might seem absurd to the reader that any sensible person could believe in the veracity of spirit and psychic phenomena. However, the gift of hindsight is often deceptive. This was a tumultuous age of transformation, in which conceptions of space and time unraveled before the eyes of Victorians. The development of railways and steamboats made it possible for Britons to travel great distances, and at speeds previously unimaginable. Equally, the construction of large-scale telegraphic networks allowed for long-distance communication at the blink of an eye. So many new technologies seemed to function through the use of invisible energies that the idea of humans possessing the capabilities of producing unseen psychic forces was not so unreasonable.[6] However, for these discussions to have any weight, reliable data to support spiritualist theories was necessary.

Psychical research, which is generally understood as the scientific study of mediumship, emerged out this context.[7] Through conversations between the journalist Edmund Rogers (1823–1910) and the physicist William Fletcher Barrett (1844–1925) during the autumn of 1881, arrangements began to be made for the formalized study of psychic forces. A conference was held a few months later, in January 1882, at the headquarters of the British National Association of Spiritualists (f. 1873). There, plans were drawn up to establish a new learned body to advance this nascent research field. It was named the Society for Psychical Research (SPR). Key areas of study within the organization included mediumship, hypnotism, disassociation and altered states of consciousness, séance phenomena, and apparitions and hauntings.[8] Thus, investigations of spirits and psychic forces were beginning to find a significant place in Victorian intellectual life, but a scientific interest in these subjects moved well beyond the activities of the SPR.

British anthropology is an ideal case study for such an investigation of modern spiritualism because issues regarding human belief had been central to the emerging discipline's research program since the 1860s. Modern spiritualism became a major preoccupation for anthropological researchers because it challenged some of the core theoretical principles of the discipline; especially the theory of animism—that primitive cultures believed that spirits animated the world, and that this belief represented the foundation of all religious paradigms.[9] In response to the issues raised by spiritualism, anthropologists turned to empirical methods as a way of upholding the veracity of their theories. As Tylor stated in his private notebook on spiritualism from 1872, "I admit a *prima facie* case on evidence, & will not deny that there may be a psychic force causing raps, movements, levitations[,] etc. but it has not proved itself by evidence of my sense."[10] What becomes clear through an examination of Victorian anthropology is that when conflicts did occur within discussions regarding spirits or psychic forces, these arguments usually revolved around issues of evidence, or lack of it, rather than around beliefs or disbeliefs per se. Therefore, I argue that when studying late Victorian anthropology's engagement with spiritualism and psychical research the emphasis should not be on its relationship to a crisis of faith, but instead to a crisis of evidence.[11] The two main concerns for most Victorian anthropologists with respect to spiritualism were determining how to become authorities in observing spiritualist activities, and how to identify credible witnesses of spirit phenomena. Understanding the observational practices of spirit and psychic investigators is therefore key to this historical study.

As a young discipline still coming to terms with its identity during the second half of the nineteenth century, British anthropology was only beginning to define its research scope. Even the language used to describe its praxis was being negotiated. Terms such as "the natural history of man," "the scientific study of man" or "race," "ethnology," "anthropology," and "ethnography" were used interchangeably—and in competing ways—depending on a researcher's disciplinary orientation, highlighting the fractured and transitional state of the emerging science. The intellectual and practicable gamut of the research field was growing too, and practitioners interested in human diversity incorporated a wide range of "visual epistemologies" into their research frameworks.[12] These methods for observing and interpreting evidence were not just about the physical act of looking at things but rather a more nuanced analytical process that involved rigorous training. Researchers working within the discipline developed all sorts of discriminating practices, which sought to identify those characteristics that were of importance to anthropologists, and those to ignore.[13] These practices included (but were not limited to) collecting instructions for researchers and informants, artistic reproduction methods for drawing, painting, and photographing subjects, cross-textual examination techniques for critically reading printed materials, and classificatory systems for arranging both natural history and anatomical and physiological evidence into manageable data sets for analyses.[14]

One of the main impediments for Victorian anthropologists was acquiring suitable data for their inquiries. Because most early ethnologists and anthropologists never left the shores of Europe, they relied on complex networks of informants for collecting their evidence. These informants came from diverse backgrounds, including colonial agents, missionaries, Indigenous cultural brokers, travelers, and medical practitioners. The structures of these collecting networks were based on the well-established systems organized in other natural history fields such as botany.[15] As was the case with botanists, many early ethnologists and anthropologists were concerned about the quality and reliability of evidence collected by informants abroad, because much of the data returning to Britain lacked the kinds of information researchers required for their studies. A key problem was that there was no systematic set of instructions for these informants to use as guidelines when collecting materials. Similar sorts of issues applied to anthropological investigations of spiritualism, which also relied heavily on the support of informants—both skeptics and believers. Making a case for or against the existence of spirits and psychic forces required substantial and trustworthy evidence, and there were well-established networks of collectors among spiritualist circles.[16]

Personal testimonies in particular were an essential source of evidence for Victorian anthropologists, contributing heavily to a growing archive that made possible the verification, expansion, and correction of ethnographic knowledge. A researcher's reputation as a trustworthy scientific investigator was based on his or her ability to either confirm the earlier observations of witnesses who had seen a particular type of cultural phenomena, or correct them based on newer information. Direct experience was essential to this process, and knowledge based on prima facie evidence added greatly to a researcher's truth-claims.[17] The ability to trust an observer was also shaped to a certain degree by their character, and "virtue epistemics" was a key part of credible witnessing. If an observer could be shown to be immoral or unethical in any capacity, it could be used to delegitimize them.[18] Anthropologists, accustomed to working with this sort of experiential data, were well placed to conduct spirit and psychic investigations, which more often than not relied solely on witness reports. The application of these observational verification techniques was easily employed in séance rooms, as well as when critically evaluating secondhand accounts in spiritualist literature. A credible witness could re-enforce his or her testimony by showing how their observations were similar to those of other witnesses professing to have seen similar phenomena. Collective empiricism was therefore a core aspect of anthropological investigations into spiritualism and psychical research during the late Victorian era.

Historiographical Context

There is surprisingly very little written on Victorian anthropology's engagement with spiritualism, particularly with regards to the movement's impact on the discipline's development. Aside from a single article by George Stocking from the early 1970s, and some passing discussions in works by Roger Luckhurst, Timothy Larsen, Jason Josephson-Storm, Elsa Richardson, and Courtenay Raia, Victorian anthropological investigations of spiritualism have been largely overlooked.[19] Christine Ferguson's formative work *Determined Spirits* (2012) comes the closest, and links the story of modern spiritualism to nineteenth-century discussions on race, eugenics, and evolution, but it is not a narrative about Victorian anthropology in the truest sense.[20] Most studies of Victorian investigations of spirits and psychics mainly trace the history through its connection to other scientific disciplines such as psychology and physics. Classic examples include the works of Janet Oppenheim, Richard Noakes, Shane McCorristine, and Peter Lamont, to name a few.[21]

There is of course a vast historiography on Victorian spiritualism, beyond studies that look specifically at its connection to modern science. Logie Barrow's *Independent Spirits* (1986), which looks at nineteenth-century plebeian culture and its engagement with modern spiritualism, was a groundbreaking book. It is, however, heavily indebted to the Marxist historiography of the time, and largely ignores scientific investigations of spirits and psychics.[22] Alex Owen's now classic book *The Darkened Room* (1989) explores the important role of women in Victorian spiritualist circles, and opened up further new and exciting areas of historical analysis. Scientific investigations (and more specifically anthropological ones) did not form an integral part of her work.[23] Literary scholars have also written extensively on the topic of Victorian spiritualism; notable examples include the works of Daniel Cottom and Helen Sword. These scholars have done much to reconsider the social, aesthetic, and political dimensions of the spiritualist movement.[24]

Nineteenth-century figures engaging in the debates over the veracity of the spirit hypothesis were very familiar with discussions occurring within science and religion on the standards of evidence. Thus, "credible witnessing" is essential to this study. At the heart of each chapter is an exploration of what it means to be a trustworthy observer of spirit and psychic phenomena. This concern has greater significance to our understanding of modern science and religion than simply illustrating some intellectual squabbles between spiritualists and skeptics in the late Victorian period. Similar debates over the reliability of witness testimonies occurred in most research fields. Three notable books to consider the importance of skilled observation for scientific practice are Steven Shapin and Simon Schaffer's seminal work *Leviathan and the Air-Pump* (1985), Tal Golan's *Laws of Men and Laws of Nature* (2004), and Lorraine Daston and Peter Galison's *Objectivity* (2007).[25] In each case these historians have complicated our understanding of what it means to observe something in a specialized way, and how witnessing and testifying was never a straightforward process, but immersed in all sorts of sociopolitical preoccupations. This process of legitimizing a person's testimony has a long tradition in "Western science." The same sorts of techniques employed during the seventeenth century by figures such as the natural philosopher Robert Boyle (1627–1691), in the rooms of the Royal Society, were reapplied by anthropologists in séance rooms during the nineteenth century. Much like other scientific pursuits, collective testimony, replication, and virtual recreation through text were all part of the arsenal for establishing the veracity of spirit investigations.[26] Credible witnessing was also important within Victorian religious circles, and biblical critics were just as skilled as their scientific counterparts in interpreting testimonies about extraordinary phenomena such as miracles. Scholars including Owen Chadwick, Victor Shea and William Whitla, Daniel Pals, and Ian Hesketh are some of the scholars to analyze this history.[27] Fundamentally, this book feeds into these larger historiographical narratives about expert witnessing in the modern age, adding further interpretive dimensions to the dialogue, especially with regard to the human sciences.

Another core historiographical problem that I explore in this book is the propensity by some scholars to view investigations into spiritualist activities as pseudoscience. Many scholarly and nonspecialist works on the rise and growth of modern spiritualism have not taken seriously the practices and theories of psychical researchers. By asking more insistently what those researchers' methods and ideas were, we can develop a more rigorous understanding of how our modern conceptions of spirits and psychic forces have been formed. Such an approach, one that firmly roots spiritualism and mediumship in the history of science, makes it possible to better contextualize the relationship between psychical research and the growth of the subfield of anthropology of religion in the nineteenth century.

In his foundational work *The Invention of Telepathy* (2002), Luckhurst argued that he wanted to explore the history of psychical research without prejudging his actors.[28] Using an historicist approach, he sought to understand his actors' interests in spirits and psychics on their own terms.[29] It might seem self-evident to pursue any historical topic in such a manner, but the historiography on modern spiritualism and psychical research is riddled with biased accounts. Spiritualist ideas and perspectives are frequently marginalized within the scholarly literature, and many accounts tend to treat modern spiritualism as credulous superstition. By giving more agency to spiritualists, there is an opportunity to create a balanced historical narrative that avoids the risk of replicating the problematic divide between the supposedly empowered and enlightened "scientific observer" and the subjugated and superstitious "observed other."[30] Spiritualists, after all, were not passive agents in these investigations, but actively contributed to the construction of knowledge about their culture. This is reflected in the way Victorian investigators often relied on spiritualists for important insights into the movement.

Trevor Hall, as an example, is representative of the more pejorative writers to tackle the history of psychical research in Britain. His books exemplify the unproblematized narrative of scientism, where rationalism conquers extraordinary belief. As a bitter ex-member of the SPR, he wrote scathing attacks on the foundational experiments of early SPR members, who were investigating spirit and psychic phenomena during the closing decades of the nineteenth century. Hall aimed to delegitimize the SPR's findings, thereby undermining the organization's research program as unscientific.[31] Other authors have gone too far in the other direction by being overly sympathetic toward spiritualist concerns, lacking any sort of recognizable critical gaze. Alan Gauld's book *Founders of Psychical Research* (1968), is a good example of this mindset, and, in many respects, his book was a counterhistory to Hall's narrative.[32]

There are also several works from the 1980s that portray a fairly whiggish narrative of the history of modern spiritualism and its relationship to science. These studies tend to begin with seemingly sensible and intelligent scientific figures being fooled into accepting the veracity of the spirit hypothesis, followed by a period of intense scrutiny and intellectual debate, before eventually arriving at a point where science pushes spiritualism and telepathy to the margins of society. Three classic examples are the writings of Janet Oppenheim, Ruth Brandon, and John Cerullo.[33] However, as Richard Noakes reminds us in *Physics and Psychics* (2019), investigations of extraordinary phenomena such as mediumship tested the boundaries and authorities of the sciences, forcing researchers to reassess and strengthen their theories and practices. The story of psychical research's supposed emergence and collapse is not a teleological narrative, but something far more complicated. Although Noakes's focus is primarily on the physical sciences, similar challenges were rampant in the human sciences as well.[34]

By bringing into conversation and emphasizing important intersections between four complementary areas of research—history of anthropology, history of scientific observation, Victorian spiritualism, and the historical relationship between science and religion—I construct a new historiographical narrative that shows how significant spirit investigations and psychical research were for late Victorian anthropology and nineteenth-century science more broadly.[35] I examine these issues by looking at the investigations of four significant anthropological figures from the late Victorian era: Alfred Russel Wallace, Edward Burnett Tylor, Andrew Lang (1844–1912), and Edward Clodd (1840–1930). I show that in their efforts to establish credible evidence either in support of or against the spirit hypothesis, they borrowed methods and theories from a wide range of disciplines, sources, and perspectives.

Vignettes, Microhistories, and Thick Descriptions

The historian of anthropology George Stocking recalled in his essay "Retrospective Prescriptive Reflections" how he struggled to "paint the big picture" during the early part of his career in the 1970s. Books, he argued, never came easy to him, and as a consequence of this shortcoming, he developed a particular historiographical method that is still widely used by historical anthropologists today. Stocking favored a microhistorical approach that examined different "vignettes" that occurred during the disciplinary history of anthropology, instead of one that provided an overarching, grand narrative. This historiographical perspective allowed him to explore what he called the "multiple contextualizations" of the discipline's past.[36] Such an historiographical method is particularly useful for a collection of case studies such as this one, because it affords an opportunity to compare how a diverse group of historical actors investigated modern spiritualism in the late Victorian era.

This book also follows in the microhistorical tradition of pioneering works by scholars such as Carlo Ginzburg, Natalie Zemon Davis, and Sigurður Gylfi Magnússon.[37] Microhistories such as these make it possible to do deep historical reconstructions of ideas and practices, which are normally impracticable in larger survey-style narratives. In figurative terms, what you see with a telescope is different from what you can see with a microscope. Not every historical actor shared the same experience, and using a microhistorical approach that examines four different figures' personal engagement with modern spiritualism allows us to see much more clearly the diverse circumstances of Victorian researchers.

Sociocultural anthropologists have also favored microstudies. For example, in his much-celebrated book *The Interpretation of Cultures* (1973), Clifford Geertz championed his "thick description" as a new way of understanding human experience. For him, anthropology focused on the "microscopic" in order to expound on larger cultural concerns. As he wrote in his opening chapter, "small facts speak to large issues," and the job of the anthropologist is to disentangle the cultural meaning from its symbolic web of ideas.[38] Whether one is studying the present or the past, a researcher should describe in detail the experiences of his subjects, contextualize them, and explain their social significance. Making sense of how ideas are formed is an exercise in "operationalism," as Geertz explained; "if you want to understand what a science is, you should look in the first instance not at its theories or its findings . . . [but] at what the practitioners of it do."[39] In the case of this book, applying a "thick description" helps us to understand the processes by which these individual historical actors formed their ideas about spirit and psychic phenomena within the context of late Victorian anthropology. Through close readings of key sources, the book unpacks the "visual epistemologies" of anthropologists investigating spiritualism, considering not only how they observed and interpreted their evidence and sources as a way of becoming "credible witnesses" but also by exposing some of the biases hidden within them.

Each of the four figures I discuss in this book represents one of the main Victorian stances on spiritualism: Wallace was a believer, Tylor was a skeptic, Lang was a revisionist, and Clodd was a disbeliever. Unpacking these varying positions can tell us a lot about the history of observation and credible witnessing in science. The four chapters that follow are designed as interconnected case studies. Each one focuses on the spirit investigations of one of these four Victorian anthropologists. For each of the chapters I focus primarily on one key text because it represents the main work that figure wrote on spiritualism. Ordered chronologically, the chapters show in much more detail how techniques for establishing the veracity of anthropological spirit investigations were refined between 1865 and 1917—the "golden age" of modern spiritualism.

In chapter 1 I situate Wallace's spiritualist writings from his most significant book on the subject, *Miracles and Modern Spiritualism* (1875), against the backdrop of Victorian anthropology.[40] I examine how Wallace, as a believer in spiritualism, constructed his argument, and how he verified the trustworthiness of his sources using theories and methods drawn from anthropology. After all, Wallace's spirit investigations have never been considered in relation to anthropology's disciplinary development in the second half of the nineteenth century. A key question was: who could be trusted as a credible witness? While much has been written on Wallace's inquiries into spirit phenomena, very little scholarship has taken seriously his remark about how his studies of spirits and mediums were a "new branch of anthropology."[41] Wallace's aim in aligning his spirit investigations to the practices of British anthropology fed into larger disciplinary discussions about the construction of reliable anthropological data. Most notably, like many of his Victorian anthropological counterparts, he grounded his research in a double commitment to firsthand observation and Baconian inductivism.

In chapter 2 I examine the skeptical observations Tylor made in his notebook on spiritualism during his trip to London in November of 1872.[42] His interest in modern spiritualism was intricately tied to his research on the anthropology of religion, and specifically his theory of animism, which he developed during the 1860s and 1870s. Animism was a theory that become one of his most important, and long-lasting, contributions to the discipline. Essentially, Tylor argued that all religions evolved from a rudimentary belief in spirits animating the world. To understand this process, one had to locate the laws that govern the development of religion, and plot all forms of worship onto an evolutionary scale, showing how religious beliefs were transformed from basic understandings of the world being animated by spirits, to complex religious systems such as Christianity. Central to Tylor's purpose was an attempt to naturalize all religions and explain their ontologies using scientific theories. He was not trying to reconcile science and religion, but to bring religion firmly within the domain of scientific understanding. Spiritualism entered these discussions because Tylor saw it as a survival of "primitive belief." Challenging the legitimacy of the spirit hypothesis became a test for demonstrating the veracity of his theory of animism, because if spirit phenomena were proven to exist, Tylor's whole argument would unravel.

In chapter 3 I discuss Lang's most significant work on the intersection of anthropology and spiritualism, *Cock Lane and Common Sense* (1894). The book is an exercise in historical reconstruction that combined ideas and approaches from anthropology, folklore, and psychology. Far from being a marginal figure from the discipline's past, Lang played a pivotal role in shaping and revising anthropological theories during the closing decades of the nineteenth century. Although he started off by following closely in the Tylorian tradition, he soon changed paths, and became far more willing to accept the reality of some unseen human faculty than his former mentor Tylor was. Therefore, *Cock Lane and Common Sense* highlighted larger anthropological debates over the veracity of animism as an explanatory theory of modern spiritualism. The repeated occurrence of ghost stories and psychic incidents over the centuries suggested to Lang that there was more to these tales than simply being superstitious remnants of bygone ages. In some respects, Lang's revisionist views on spirits and psychic forces had much more in common with Wallace's ideas in *Miracles and Modern Spiritualism* than with Tylor's in *Primitive Culture* (1871).[43]

In chapter 4 I explore the hardline disbelief of Clodd in his book *The Question: A Brief History and Examination of Modern Spiritualism* (1917).[44] At its core, the book is a comprehensive study of the types of evidence used in debates over the existence of spirits and psychic forces. Clodd traced the historical development of the spirit hypothesis, discussed advances in psychical research, looked at the careers of prominent mediums, and classified various types of so-called spirit phenomena using the taxonomic systems of Victorian anthropologists. The evidence, from his perspective, was overwhelmingly against the view that spirits and psychic forces were real, and he concluded that on the whole, modern spiritualism was the product of deception, credulity, and superstition. Fundamentally, Clodd argued that modern spiritualism was a sham, and that the veracity of the theory of animism for explicating the reasons for belief in spirits and psychics was sound. His book was one of the fiercest critiques of modern spiritualism forwarded by an anthropologist, and Clodd hoped that his work would finally put an end to the spiritualist movement.

Focusing on anthropology's engagement with modern spiritualism exposes an important story in the history of the discipline's development during the second half of the nineteenth century, and the significance of scientific observation and visual epistemologies in this process. Traditionally, late Victorian anthropology has been framed as an armchair pursuit, committed to deductive reasoning, with little to no firsthand experience engaging directly with different cultures. Yet the discipline's engagement with modern spiritualism tells a different story, one that emphasizes the importance of inductive methods in late Victorian anthropological research. What will become clear is that anthropologists have always been preoccupied with the ways in which evidence is made credible, and the types of theories and practices that determine the accuracy of their information.

The historical relationship between science and religion is also brought to the fore through this examination of anthropology's engagement with modern spiritualism. The very decision by figures such as Wallace, Tylor, Lang, and Clodd to investigate spiritualists in the first place was linked to larger societal discussions on the so-called crisis of faith and secularization in the nineteenth century. And yet, although the historical actors' discussions may have started off by focusing on issues relating to belief or disbelief in spirits or psychic forces, the ensuing debates that occurred through the investigations of mediums and spiritualist performances centered on issues of evidence, or lack of it. Therefore, the crisis of evidence took center stage, and empiricism was a means to an end for finally resolving the spirit hypothesis conundrum.

1

Alfred Russel Wallace

The Believer

Alfred Russel Wallace—the celebrated codiscoverer of evolution by natural selection—was a self-proclaimed skeptic of spiritualism before attending his first séance on July 22, 1865, at the home of his friend Lewis James Leslie (b. 1806) in Tunbridge Wells, Kent.[1] "Up to the time when I first became acquainted with the facts of Spiritualism," Wallace wrote in the opening pages of *Miracles and Modern Spiritualism*, "I was a confirmed philosophical sceptic. . . . I was so thorough and confirmed a materialist that I could not at that time find a place in my mind for the conception of spiritual existence, or for any other agencies in the universe than matter and force."[2] Like most men of science during the Victorian age, he did not believe that there was sufficient evidence available to confirm that spirits or psychic forces were real.[3]

This event in 1865, however, proved to be a transformative moment in Wallace's life, and his stance began to change. Over the course of the next year, he immersed himself in the study of the supernatural, becoming a vocal proponent of modern spiritualism.[4] On November 22, 1866, Wallace wrote a letter to his friend the biologist Thomas Henry Huxley (1825–1895) about an exciting "new branch of anthropology" that he had been working on—the investigation of spirit phenomena. Wallace invited Huxley to join him on a Friday evening so that he too could witness the curious displays and perhaps contribute to this budding field of scientific inquiry.[5] Huxley quickly rejected the invitation, stating, "I have half-a-dozen investigations of infinitely greater interest to me which any spare time I may have will be devoted."[6] Wallace's exchange with Huxley foregrounds how difficult it was for proponents of spirit investigations to gain scientific credibility.

Wallace believed that if he could go from being a skeptic to a convert, so too could other men of science. Arguing that the key to achieving scientific legitimacy for spirit investigations was through the establishment of reliable witnesses and trustworthy evidence, he committed himself to that cause for the remainder of his life.[7] As Janet Oppenheim argues, Wallace "was convinced that the testimony of his own sense, combined with the records of countless other investigators over the centuries, provided an adequate empirical base on which to establish the validity of spiritualism."[8] To verify his belief, Wallace consciously borrowed observational techniques from anthropology—a science that relied quite heavily on credible witnessing.[9] What it meant to observe something anthropologically, for Wallace, was not simply the physical act of looking at something; rather, it was a much more nuanced process of collecting, analyzing, and representing anthropological data.[10] It is a way of knowing and understanding the world through a specialized framework—what Daniela Bleichmar calls "visual epistemology."[11]

While much has been written on Wallace's inquiries into spirit phenomena, very little scholarship has taken seriously his remark about how his studies of spirits and psychic forces were a type of anthropology.[12] When connections are drawn between Wallace's twin interests in spiritualism and anthropology, it is usually to examine his research on extra-European conceptions of spiritualism.[13] While there is no doubt that many of the practices that Wallace used in his spirit investigations were shared by other sciences, most notably physics and natural history, it is telling that he identified anthropology as being the best fit for developing his research program. It was fairly common during the middle of the nineteenth century for anthropologists to borrow techniques from the more established sciences. Physics, geology, and natural history provided a means to an end for achieving more authority within the larger scientific community. Nevertheless, when methods from the physical and natural sciences were employed in Wallace's spirit investigations, they were typically reframed along anthropological lines.

Wallace's aim of aligning his spirit investigations to the practices of British anthropology fed into larger disciplinary discussions about the construction of reliable anthropological data. Most notably, Wallace—like many of his Victorian anthropological counterparts—grounded his research in a double commitment to firsthand observation and Baconian inductivism. His insistence on "fact-based," experiential knowledge echoed the disciplinary rhetoric of the physician and speech therapist James Hunt (1833–1869), who cofounded the Anthropological Society of London (ASL) in 1863, along with the orientalist, explorer, and military officer Richard Francis Burton (1821–1890).[14] It is hardly surprising that many of the methods and theories that Wallace imposed on his studies of spirit phenomena included core aspects of Hunt's anthropological vision. After all, Wallace was conducting much of the research for *Miracles and Modern Spiritualism* while he was regularly attending ASL meetings during the height of the anthropological schism of the 1860s. His spirit investigations were written at a time when he was engrossed in the debates and discussions on how to build a new anthropological science in Britain.[15]

This was a key period in the disciplinary history of British anthropology, in which two rival groups of researchers were vying for control of the race sciences.[16] On one side were the anthropologists led by Hunt, who, generally speaking, promoted a form of polygenesis that was grounded in biological determinism, and emphasized directly observable "facts" to support its suppositions. On the other side were the ethnologists led by Huxley at the Ethnological Society of London (ESL), who promoted a form of monogenesis that was grounded in a mixture of older Prichardian historicism and newer Darwinian evolutionary theory.[17] During the course of this dispute—which raged between 1863 and 1871—both sides published several essays in the pages of their societies' journals on the scope of their respective research fields. Although both camps claimed to be doing distinct forms of scientific investigation, there were many overlapping topics, theories, and practices in their respective research programs.[18] Even the membership lists of the two societies contained many of the same names. The focuses of anthropology and ethnology during this period were quite broad, and for the most part comprised any study that examined the cultural and physical aspects of human groups. Eventually, after several attempts to reconcile the grievance, an agreement was reached, and in 1871 the two societies merged to form the Royal Anthropological Institute of Great Britain and Ireland (RAI). This new learned body brought together the methods and theories of researchers from both camps.[19]

Underlying many of these disciplinary debates were discussions on how to observe something anthropologically or ethnologically, and this preoccupation maps onto Wallace's research into modern spiritualism. The ability to observe spirit phenomena as a credible witness was a central concern for Wallace. In many respects it underscores the primary objective of his *Miracles and Modern Spiritualism*. Like his peers at the ASL, who were striving for proper recognition within the British scientific world, Wallace, too, was fighting for the scientific legitimacy of his spirit investigations. Anthropology provided a tactical blueprint for achieving this status. If Wallace were to prove scientifically that humans transcended into spirits upon death, then anthropology was the ideal discipline to anchor his spirit investigations in. After all, figures such as Hunt purported that anthropology was the only discipline to examine all aspects of human life, and for Wallace that would include the afterlife.[20] By bringing spiritualism into anthropology, the discipline would be justly able to argue that it studied the "entirety" of human existence.

Alfred Russel Wallace and the Making of a Spiritualist and Anthropologist

Wallace was born in 1823 in the Welsh village of Llanbadoc near Usk in Monmouthshire. He was the eighth child of Mary Anne (1792–1868) and Thomas Wallace (1771–1843). His father had trained in law but never practiced, and had several unsuccessful businesses. At the age of five, Wallace and his family moved to Hertfordshire, where he attended Hertford Grammar School, until family financial difficulties forced him to withdraw at the age of fourteen.[21] He spent the next few months living in London with his older brother John (1818–1895), who was a builder. While living in London, Wallace regularly visited the London Mechanics Institute (f. 1823), where he began reading a range of scientific works. It was also during this period that he was exposed to the political ideas of the Welsh social reformer Robert Owen (1771–1858). Early into his London stay, Wallace began attending political lectures at the Owenite Hall of Science near Tottenham Court.[22] His participation in these political activities had a long-lasting impact on his beliefs, including his views on spiritualism.

Core to Owenism were communitarian values including universal suffrage, unconventional religious worship, cooperative labor, and moral instruction. Owen himself became a believer of the spirit hypothesis in 1853 after attending séances with the American medium Maria B. Hayden, who is widely regarded as the first modern medium to practice professional spiritualism in England. Owen outlined his new spiritual beliefs in a pamphlet titled *The Future of the Human Race* (1854).[23] Wallace's later socialistic ideas about spiritualism were indebted to his early engagement with Owenism, and he championed a more utopian understanding of humanity in the afterlife. As we will see in due course, Wallace contended that through the spiritual evolutionary process, which began upon death, all spirits could achieve a form of enlightenment through moral education.[24]

Eventually, Wallace left London in search of more stable employment opportunities. He spent nine months apprenticing as a watchmaker before moving to Kingston, Herefordshire, in 1839 to work for his other brother William (1809–1845), who owned a land-surveying business. Britain's railway industry was booming after the Tithe Commutation Act was passed in 1836, and there was plenty of work to be found as a surveyor. Under the tutelage of his brother, Wallace gained a broad practical skillset in drafting, mapping, mathematics, and architectural design and construction. This practical knowledge helped to establish Wallace as a credible and scientifically minded researcher, which would later serve him well in his career as a traveling naturalist and specimen collector. Moreover, these early experiences ultimately contributed to his reputation both as a respected anthropologist and ethnographer and as a reliable investigator of extraordinary phenomena at séances and other spiritualist performances.[25]

Although it is reasonable to argue that Wallace's investigation of modern spiritualism encompassed far more than what is typically viewed as anthropological theory and practice, it remains rather telling that he framed his study of spirits and psychic forces as a "new branch of anthropology."[26] There was much overlap between the newly emerging disciplines during the middle of the nineteenth century, and to fully understand why Wallace viewed his research as a type of anthropology, as opposed to physics, biology, or natural history, it is necessary to map the core aspects of his spirit investigations against the key criteria for a study to be considered "anthropological" during the mid-Victorian period. These criteria can be synthesized in a rudimentary fashion as follows: first, it had to focus on humans (broadly construed); second, there should be some elements of historicism or evolutionism in it; third, it should use the Baconian method of induction; and fourth, it should be derived from experiential knowledge, based on directly observable "facts." Wallace's spirit investigations tick each of these boxes.[27]

Wallace's double interests in spiritualism and anthropology developed over twenty-five years, and tracing this process allows for a more sophisticated understanding of Wallace's spirit investigations. Although he credited his first séance in 1865 as a transformative moment in redefining his views on spiritualism, Wallace had encountered what he believed to be supernatural phenomena much earlier in his life.[28] For example, in 1843, while working as a surveyor in rural Wales, Wallace wrote a short ethnographic account on the local Welsh farmers, with considerable attention paid to their customs, habits, and traditional folkloric beliefs. He remarked, for instance, that the farmers were "exceedingly superstitious," and that "witches and wizards . . . are firmly believed in, and their powers much dreaded."[29] He also described other types of folkloric practices among the Welsh farming community. The "corpse candle," for instance, was an object for the customary ritual of carrying "a lighted candle, which is supposed to foretell death . . . from the house in which the person dies along the road where the coffin will be carried to the place of burial."[30] The local farmers ascribed much symbolic supernatural meaning to this performance, and as Wallace recounted, the community often viewed its practice as a harbinger of further death. These early ethnographic experiences in Wales were significant for Wallace, sensitizing him to a belief in the potential existence of extraordinary phenomena, and enriching his knowledge of fieldwork methods. Moreover, they represented an important stepping-stone for him in recognizing the value of direct observation in the study of alleged supernaturalism. This high valuation of prima facie evidence would become a benchmark of Wallace's later writings on spirits and psychic forces.

Another of Wallace's early important experiences with extraordinary forces occurred in 1844, after he left his job as a surveyor to become a schoolteacher in Leicester. It was there that Wallace witnessed the curious effects of mesmerism during a lecture delivered by the renowned phrenologist and mesmerist Spencer Timothy Hall (1812–1885).[31] The spectacles that he had observed intrigued him, and Wallace set about conducting some rudimentary experiments with mesmerism on his students. He later recalled in *Miracles and Modern Spiritualism* that he was able to influence some of his students using techniques similar to those employed by Hall in his lecture.[32] These alleged positive results had a lasting impression on him, and, as Martin Fichman has argued, Wallace's early research on mesmeric forces "predisposed him to remain open to claims relating to psychic phenomena." Nevertheless, he was a skeptic for the time being.[33] Other incidents, such as near-death experiences during his travels to both South America between 1848 and 1852 and the Malay Archipelago between 1854 and 1862, when he was working as a natural history specimen collector, further exposed Wallace to what he believed to be supernaturalism. For example, Wallace stated, "At least three times within the last twenty-five years I have had to face death as imminent or probable within a few hours, and what I felt on those occasions was at most a gentle melancholy at the thought of quitting this wonderful and beautiful earth to enter on a sleep which might know no waking. In a state of ordinary health I did not feel even this. I knew that the great problem of conscious existence was one beyond man's grasp, and this fact alone gave some hope that existence might be independent of the organized body."[34] According to Wallace, nearly dying on these three occasions sensitized him to a feeling of a deeper existence beyond the mortal world—one that would be essential to his later "theory of spiritualism."[35]

It was also during Wallace's travels that the modern spiritualist movement arose. The first high-profile mediums were the American Fox sisters, Leah (1831–1890), Margaret (1833–1893), and Kate (1837–1892). In 1848, while the siblings were living with their parents in Hydesville, New York, the two younger sisters, Margaret and Kate, allegedly began communicating with spirits through rapping.[36] News of these spirit communications spread quickly across the United States, Britain, and continental Europe. Within a few years more mediums were coming to prominence through private and public performances. A proliferation of spiritualist literature was also being published, such as the *Yorkshire Spiritual Telegraph*, which was the first modern spiritualist periodical to be produced in Britain in 1855. Many popular lecturers on spiritualism and mediumship also toured North America and Europe during this period. For example, the young trance medium Cora L. V. Scott (1840–1923) was an early spiritualist to gain some notoriety by delivering public presentations to broad audiences during the early 1850s in the northeastern United States.[37] Around the same time, Hayden was building a devout audience in London. She was soon followed by another American medium to visit the British capital, Mrs. Roberts. Unlike Hayden, who was primarily known for producing spirit rapping during her performances, Roberts's specialty was table turning.[38] Thus, the British populace was being introduced to a range of supposed extraordinary phenomena through these newly emerging spiritualist performers, and during the 1850s proponents of spiritualism were clearly on the ascent.[39] Wallace had been hearing about this social phenomenon while traveling abroad, and according to Fichman he was determined to investigate the matter himself upon returning to England.[40]

While traveling through South America and the Malay Archipelago, Wallace further developed a strong interest in ethnography and ethnology. Like most European scientific travelers of the nineteenth century, Wallace immersed himself in travel literature, particularly accounts written by Europeans who had visited the same regions he did. Here Wallace was following a practice of informed reading that was an important preparatory exercise for journeys into relatively unknown lands.[41] It provided travelers with essential information on the peoples, plants, animals, and landscapes they were going to confront abroad. This was especially useful for natural history collecting, which remained the main source of Wallace's income during the period. Knowledge of these works made it possible for someone such as Wallace to establish himself as a credible observer of natural history data.[42] It was also through traveling that Wallace's intersecting interests in ethnography and spiritualism transformed. His cross-cultural encounters with Indigenous peoples exposed him to various forms of spiritualism that would later influence his spiritualist beliefs, and as Sherrie Lynne Lyons has argued, "Wallace's view of native people provides an important clue to his later conversion to the spiritualist hypothesis."[43]

For many early ethnologists, travel narratives provided a crucial source of data for their studies. These narratives contributed to a growing archive that made possible the verification, expansion, and correction of ethnographic knowledge. One's reputation as a reliable scientific traveler was based on the ability to either confirm the earlier observations of travelers who had visited the region or correct those reports based on newer information. In both cases, direct experience was essential to this process.[44] Being able to state that the data were acquired firsthand added greatly to one's truth-claims. Impartiality was also an essential element of this process, and in many respects the method of verifying ethnographic observations was closely linked to what Lorraine Daston and Peter Galison have termed "truth-to-nature objectivity," in which the accepted representation of a human group was a compilation of reports, or the archetype of a pattern within various accounts.[45] When substantiating the trustworthiness of his observations of spirit phenomena, or those of other witnesses that he deemed credible, Wallace used similar ethnographic techniques. Although not explicitly mentioned in his writings, much of Wallace's analysis was influenced by the works of the German naturalist and traveler Alexander von Humboldt (1769–1859), who combined Baconian and Linnaean principles with his own ideas about how scientific travelers could systematically catalogue the natural world. Humboldt argued for a *physique du monde*—a universal science based on observational study, measurement, and experimentation.[46]

The verification of direct observations through ethnographic methods was a core aspect of Wallace's later anthropological writings on spiritualism. These investigations were further enhanced by Wallace's knowledge of ethnological and anthropological theories. For instance, Wallace had read major ethnological works by the physicians James Cowles Prichard (1786–1848) and William Lawrence (1783–1867) while traveling throughout the world. Both men were foundational figures within the discipline, and outlined theoretical frameworks for studying human races.[47] Once Wallace returned to Britain in 1862, he joined the ESL, and when the ASL formed a year later, Wallace also began attending their meetings. This further immersed Wallace in discussions on how to do ethnological and anthropological research. His first major contribution to anthropological studies occurred in 1864 when he published what is widely considered to be the first anthropological work in Britain to apply Darwinian evolutionary mechanisms to the study of humans.[48] It was clear that in the early 1860s much of Wallace's scientific activities were grounded in ethnological and anthropological research. Thus, by the time Wallace attended his first séance in July 1865, he was primed for approaching the event as a skilled ethnographic observer. In many respects his later descriptions of his experiences at séances can be seen as a kind of ethnographic reporting, informed by his deep knowledge of ethnological and anthropological theories.

The "Theory of Spiritualism" and Evolutionism

Miracles and Modern Spiritualism contained three substantially reworked versions of essays that Wallace had produced between 1866 and 1874.[49] As he immersed himself in the modern spiritualist movement, his knowledge deepened, and he was able to incorporate even more information from his readings of the extant spiritualist literature, and from his personal experiences at séances. Thus, the book represents the culmination and maturation of his early spirit investigations. In his book, Wallace recognized that the nature of his investigations meant that many of his readers would be skeptical of the genuineness of the phenomena he described, and he wrote, "Many of my readers will, no doubt, feel oppressed by the strange and apparently supernatural phenomena here brought before their notice. They will demand that, if indeed they are to be accepted as facts, it must be shown that they form a part of the system of the universe, or at least range themselves under some plausible hypothesis."[50] It was for this reason that he carefully articulated a "theory of spiritualism."[51] Through the conceptualization of his theory, Wallace believed that he was strengthening the scientific pronouncements of his research. Crucial to this process was grounding his spirit investigations in naturalistic laws, and providing conceivable explanations for psychic forces.[52] This enabled Wallace to align his spiritualism with the core principles of scientific naturalism that were dominant during the period. It was a tactic that other ethnologists and anthropologists were using as they strove for recognition in the larger Victorian scientific scene. Wallace was following this approach. However, the version of scientific naturalism that he employed differed from the more biologically determined model of Hunt, and the Darwinian-inspired model of Huxley and other X Club members. Wallace was far less committed to hardline verificationist assumptions, allowing him to be more receptive to extraordinary phenomena, such as spirits and psychic forces.[53]

At the crux of Wallace's theory was a fundamental principle that underlay all spiritualist phenomena—every human was made of two parts: the spirit and the material body. Wallace believed that the material body was the "machinery and instruments by means of which [humans] . . . act upon other beings and on matter," and the spirit "feels, and perceives, and thinks."[54] While the material body would eventually perish, the spirit was immortal. Once the spirit entered the afterlife, it began a developmental process, which Wallace termed "progression of the fittest."[55] It was a new form of human developmentalism that attempted to reconcile his evolutionary and anthropological ideas with his spiritualist ones. This merger of ideas firmly posited Wallace's spirit investigations in larger discussions on the epistemic limits of science and religion. Although many midcentury scientific naturalists were attempting to displace religion and remove it from scientific studies, Wallace's position differed, and he saw the two spheres as being compatible so long as they were employed through a spiritualist framework. It was a much different approach from the two spheres model favored by other scientific naturalists such as Huxley, the philosopher of science Herbert Spencer (1820–1903), and the physicist and lecturer John Tyndall (1820–1893). Thus, Wallace's spirit investigations complicate our historical understanding of the relationship between science and religion, exposing the fluidity of the supposed boundaries.[56]

Wallace wanted to map his theory of spiritualism onto his own evolutionary paradigm. The inclusion of evolutionism in ethnological and anthropological research was essential for many scientific naturalists during the 1860s and 1870s. Figures such as Huxley, who was president of the ESL during the 1860s; the archaeologist, entomologist, and politician John Lubbock (1834–1913), who was the first president of the RAI; and Edward Burnett Tylor, who is widely regarded as the founder of cultural anthropology and the first researcher in Britain to be appointed as reader in anthropology, all trumpeted the importance of evolutionary theories for the advancement of the discipline.[57] Wallace was part of this general movement in anthropology, and it was an underlying theme in his first major contribution to anthropological research: his 1864 article "On the Origin of Human Races and the Antiquity Man Deduced from the Theory of Natural Selection."[58] Given that Wallace viewed the study of spirits and psychical forces as a type of anthropological pursuit, evolutionism also formed a key aspect of his strategy for ensuring that his spirit investigations gained acceptance within scientific circles.[59]

For Wallace, nondirectional evolutionary processes, guided by natural selection, may have accounted for human diversity when studying the living, but upon death a different sort of evolutionary process began that was directional and progressive. He wrote, "The organic world has been carried on to a high state of development, and has been ever kept in harmony with the forces of external nature, by the grand law of 'survival of the fittest' acting upon ever varying organisations. In the spiritual world, the law of the 'progression of the fittest' takes its place, and carries on in unbroken continuity."[60] According to Wallace the spirit was a mind without body that retained all the knowledge (both intellectual and moral) it had acquired during life, including the experiences, thoughts, feelings, and tastes of the former self.[61] With this knowledge the spirit could progress through successive stages toward the highest level of enlightenment. Spiritual growth was humanity's true purpose, and the influence of Wallace's early interests in Robert Owen's conception of moral education is clearly present in the framework.[62] However, unlike Owenism, which was largely deistic, Wallace's spiritual framework was theistic, and as Fichman has observed, it "explicitly maintains that the Divine Being continues to sustain relations to His creation" even after death.[63] Because the spirit retained knowledge of its former self, Wallace argued, it was able to communicate with the living. How else would it be possible for spirits to verify their identities to living loved ones during séances?[64]

Wallace's inclusion of a developmental model into his theory of spiritualism was an essential part of his aim to gain scientific legitimacy for his spirit investigations. Because it combined familiar aspects of Darwinian evolution with monogenesis and progressivism, his theory of spiritualism could be incorporated into larger ethnological and anthropological discussions about human evolution. However, it moved beyond these discussions to new ground, and unlike the standard forms of monogenesis (or even polygenesis, for that matter) it provided an alternative framework that accounted for the evolution not only of the living but also of the dead. Nevertheless, in order to support this spiritualist evolutionary paradigm, Wallace required data that proved the existence of spirits and psychic forces. This led him to emphasize "fact-based" knowledge and Baconian inductivism, further tying his spirit investigations to the techniques of ethnologists and anthropologists in the mid-Victorian period.

Direct Observation and "Fact-Based" Knowledge

When Hunt cofounded the ASL in 1863, he wanted anthropologists to distinguish themselves from ethnologists by prioritizing research that was based on direct observation and "fact-based" knowledge. One of the main criticisms that his opponents such as Huxley and Lubbock leveled at him was that it was unnecessary for there to be two disciplines that essentially studied the same materials. Hunt had to rationalize the formation of anthropology as distinctly different from ethnology. To substantiate the scientific criteria on which anthropologists based their analyses, he argued for the application of the Baconian method of induction.[65] Using Baconianism to strengthen the scientific pronouncements of a discipline was by no means distinct to anthropology, and many emerging disciplines in the nineteenth century appealed to Baconian principles as a way of gaining scientific legitimacy within the larger community.[66] It was primarily a rhetorical strategy, in which the Baconian method of induction was championed as a means to knowledge. More often than not, though, its application was never fully implemented in research programs.[67]

In the case of anthropology, Hunt asserted that Baconianism, with its emphasis on "facts," was the most reliable way to do scientific research. He wrote, "It has been solely the application of this [Baconian] method which has given such weight to our deliberations and our deductions. Loyalty to facts with regard to . . . anthropology brought us face to face with popular assumptions, and the contest has resulted in victory to those who used the right method."[68] Under this model, Hunt strategically used the Baconian method of induction as the cornerstone of his anthropological framework. Observable "facts" such as anatomical and physiological data lay at the foundation of any good study of humans. There were three steps to using the Baconian method of induction. First, researchers were to collect materials and describe the facts. Second, they were to tabulate or classify the facts into three categories: a) instances where a specific characteristic was observed, b) instances where the characteristic was absent, and c) instances where a variation of the observed characteristic was present. Third, in light of what the tabulated materials demonstrated, researchers were to draw conclusions based on the data and determine which phenomena (physical or cultural) were connected to it, and which phenomena were not.

Even within the human sciences Hunt's emphasis on "fact-based" knowledge was not distinct, and despite his claims to the contrary, ethnologists such Prichard and Lawrence had been using the Baconian method in their research programs for several decades prior to the emergence of anthropology. The only difference was that Hunt explicitly identified its importance because he was attempting to legitimize his methodological approach as more rigorous than the practices of the researchers who preceded him.[69] Tylor also used a similar "fact-based" argument in the opening pages of his magnum opus, *Primitive Culture* (1871). Tylor was constructing an evolutionary model that traced the development of cultural attributes from what he believed to be the lowest stages of human societies to the most advanced. Because he was primarily working with nonphysical aspects of cultures, he argued that it was essential to stockpile his "facts," and show multiple examples of similar cultural phenomena. Such an approach would verify the trustworthiness of his evidence and support his suppositions. Tylor stated, "Should it seem to any readers that my attempt to reach this limit sometimes leads to the heaping up of too cumbrous detail, I would point out that the theoretical novelty as well as the practical importance of many of the issues raised make it most unadvisable to stint them of their full evidence."[70]

There is a similar sort of rhetoric in Wallace's *Miracles and Modern Spiritualism*. Like Hunt and Tylor, Wallace strategically argued that spirit investigations had to be grounded in firsthand knowledge and reliable "facts." It was one of the many ways that their anthropological research intersected. He placed great emphasis on direct observation: "In this manner only could all sources of error be eliminated, and a doctrine of such overwhelming importance be established as truth. I propose now to inquire whether such proof has been given, and whether the evidence is attainable by any one who may wish to investigate the subject in the only manner by which truth can be reached—by direct observation and experiment."[71] Wallace's visual epistemology, to borrow the term from Bleichmar, began with firsthand experience, and for him seeing was knowing.[72]

Direct observation was only part of establishing the credibility of one's investigations. A single observation could be fabricated, erroneous, or accidental. Therefore, it had to be compared to similar reports. Wallace wrote, "A single new and strange fact is, on its first announcement, often treated as a miracle, and not believed because it is contrary to the hitherto observed order of nature."[73] It was for this reason that Wallace insisted on finding multiple examples of similar accounts to verify the credibility of a single observation. Their aggregation was testament to genuine occurrences. If one of the "facts" from a collection of observations were proven to be real, the rest by extension would also be taken as factual. Thus, it was a circular verification process. Wallace argued, "If but one or two of them are proved to be real, the whole argument against the rest, of 'impossibility' and 'reversal of the laws of nature,' falls to the ground."[74]

The influence of Baconianism is noticeable in other ways in Wallace's *Miracles and Modern Spiritualism*. As was the case with Hunt and others, Wallace organized his facts into groupings for analysis. For example, he sorted physical and mental spirit phenomena into different taxa. Each grouping included short descriptions to assist researchers in correctly identifying the types of phenomena that they were witnessing. Wallace believed that he was articulating a system that allowed for more accurate observations and a clearer understanding of the different kinds of evidence available. If every spirit investigator used his system in their studies, it would create a standardized method, which in turn would make it easier to compare reports. The replication of results would also be possible under such a model, which added further authority to the method.[75]

Under "physical phenomena types" Wallace included six categories: "simple physical phenomena," such as sounds being produced without a known source, or people being moved without any human agency involved; "chemical," where mediums could hold burning-hot objects without getting hurt; "direct writing and drawing," which included pencils and pens allegedly rising on their own; "musical phenomena," where instruments played without any human intervention; "spiritual forms," such as the appearance of ghosts, orbs, or specters; and finally "spirit photographs," which purported to include manifestations of spirits in them.[76] There were five types of "mental phenomena" described in Wallace's system, and they included "automatic writing," when mediums wrote information down involuntarily; "seeing or clairvoyance," such as premonitions or voice hearing; "trance speaking," when a medium communicated the thoughts or feelings of a spirit; "impersonation," which is closely linked to trance speaking, but is more about speaking or acting like the spirit; and finally "healing," such as a medium detecting an unknown illness in a person.[77]

Whereas physical phenomena did not require a medium to occur, mental phenomena did. It was for this reason that Wallace argued that mental phenomena were usually considered less evidential by skeptics of spiritualism. He wrote, "The purely mental phenomena are generally of no use as evidence to non-spiritualists, except in those few cases where rigid tests can be applied." Wallace continued by arguing that the two kinds cannot be separated so easily and that "they are so intimately connected with the physical series, and often so interwoven with them, that no one who has sufficient experience to satisfy him of the reality of the former, fails to see that the latter form part of the general system, and are dependent on the same agencies."[78] For Wallace, proving the existence of spirit forces meant using all forms of data, and seeing how the sum of the parts came together to show the reality of spiritualism.

Wallace's commitment to direct observation and "fact-based" knowledge was essential for establishing the credibility of his spirit investigations. Like

Hunt and Tylor before him, he had to justify his "new branch of anthropology" as a legitimate scientific pursuit. It was not enough to have a comprehensive theory of spiritualism and a few examples of direct observations that purported to show genuine examples of spirit phenomena. Wallace asserted that proponents of spiritualism had to show that the sheer volume of credible witnesses collecting data on spirit forces was impossible to ignore, and therefore had to be taken seriously. He stated,

> I maintain that the facts have now been proved, in the only way in which facts are capable of being proved—viz., by the concurrent testimony of honest, impartial, and careful observers. Most of the facts are capable of being tested by any earnest inquirer. They have withstood the ordeal of ridicule and of rigid scrutiny for twenty-six years, during which their adherents have year by year steadily increased, including men of every rank and station, of every class of mind, and of every degree of talent; while not a single individual who has yet devoted himself to a detailed examination of these facts, has denied their reality. These are characteristics of a new truth, not of a delusion or imposture. The facts therefore are proved.[79]

In sum, the success of Wallace's methodological approach to achieving scientific legitimacy for his spirit investigations rested on the trustworthiness of his sources. The bulk of Wallace's *Miracles and Modern Spiritualism* is devoted to establishing the credibility of his eyewitnesses and the reality of the phenomena. Once again, ethnological and anthropological theories and methods were essential to this process.

Credible Witnessing in Wallace's Spirit Investigations

Ethnographic reports from the narratives of travelers were the modus operandi of ethnological and anthropological research during the nineteenth century. Because most practitioners never left the shores of Europe, they were reliant on travelers for their data. Narratives provided essential information on the peoples who inhabited the world, and the personal testimony of someone who had seen different races directly in situ always had more authority than secondhand descriptions based on a priori knowledge. Ethnographies of Europe used similar techniques—especially when verifying personal testimonies from historical periods. Thus, there was a long tradition in both ethnology and anthropology of establishing the credibility of firsthand observers. Figures such as Prichard, Lawrence, Robert Gordon Latham (1812–1888), Hunt, and Tylor all worked tirelessly to prove the accuracy and trustworthiness of their sources.[80] Wallace was building on this tradition in his spirit investigations, and he had a broad knowledge of both ethnological and anthropological methods from his participation at both the ESL and ASL during the 1860s.[81]

There were several ways of establishing the reliability of a source in ethnology and anthropology. If the observer possessed comprehensive training in a field that was considered to be requisite to ethnology or anthropology, such as medicine or natural history, they were considered to be a credible witness. Similarly, if an observer had a sound knowledge of subjects such as law, physics, or philosophy, they were also deemed trustworthy because of their analytical and discerning mind. If multiple accounts contained analogous information on the same objects, topics, events, or peoples, they collectively reinforced the validity of one another. It was a sort of "collective empiricism"—to borrow a term from Daston and Galison.[82] Any inconsistency that appeared within the dataset would be identified as atypical and subsequently removed. If a researcher could reinforce the claims of other observers through their own similar firsthand experiences, this added further credibility to an account. Finally, if multiple witnesses were present at the same incident and produced corresponding reports, they too were deemed trustworthy observers. We see examples of all of these modes of verification in Wallace's spiritualist writings.[83]

A reliance on personal testimonies to substantiate one's scientific suppositions was of course a tried and trusted method in most scientific disciplines. As Steven Shapin and Simon Schaffer have shown in their work, since the early modern period practitioners in various fields regularly appealed to different forms of collective empiricism to support their research activities.[84] However, Wallace gave particular credence to researchers working in the life sciences, because he believed that they were less predisposed to imposing set conditions onto the study of unexplained phenomena.[85] When it came to observing spirit activity, physicists could make important and valuable observations, but the analysis of these reports was left to scientists with more detailed understandings of the organic world. In a sense, we can think of this method as a two-part process. First, information had to be observed and recorded by credible witnesses. Second, the meaning of the phenomena that were witnessed in the reports had to be interpreted and explained by a researcher with a strong grounding in natural history and the human sciences. Because experiential knowledge of human societies was the backbone of ethnological and anthropological research, its practitioners were particularly well suited for synthesizing and interpreting evidentiary materials of these kinds. This approach to making sense of observational accounts, which Wallace was employing in his spiritualist writings, was a staple of early armchair-based anthropological research. Wallace saw himself as the quality controller of the data, weeding out any anomalous evidence, and highlighting examples that best supported his suppositions. [86]

In *Miracles and Modern Spiritualism*, Wallace made a case for the credibility of his sources. He argued that despite the number of skeptics of spiritualism rising over a twenty-year period (between the 1850s and 1870s), the number of believers was still growing.[87] He asserted that the kinds of data, and the rigor of experimentation purporting to prove the existence of spirits and psychic forces, were improving, and much of this was the result of a general increase in the number and quality of reports that were produced by so-called respectable observers of spirit phenomena. Wallace stated, "I shall call chiefly persons connected with science, art, or literature, and whose intelligence and truthfulness in narrating their own observations are above suspicion; and I would particularly insist, that no objections of a general kind can have any weight against direct evidence to special facts, many of which are of such a nature that there is absolutely no choice between believing that they did occur, or imputing to all who declare they witnessed them, willful and purposeless falsehood."[88] Wallace's avowal of the credibility of his sources was grounded in three key points: First, all of his witnesses were leaders in science, art, or literature, and therefore represented some of the greatest minds known. Second, the reports were directly observed, and were not based on a priori assumptions. Wallace argued that this meant they were stronger sources of evidence. Third, because the information was acquired through firsthand experience, all of the observations were founded in "fact-based" knowledge. As sources of data, Wallace believed they should be considered as of the highest caliber.

The pages of *Miracles and Modern Spiritualism* are littered with copious observations recorded by figures whom Wallace deemed credible witnesses of spirit phenomena. He was particularly favorable toward reports that described séances led by eminent mediums such as Kate Fox, Daniel Dunglas

Home (1833–1886), and Agnes Elisabeth Guppy (1838–1917). Each of these mediums was celebrated by spiritualists as possessing genuine psychic powers, and therefore attracted the attention of many high-profile spirit investigators (both skeptics and believers). There had been multiple attempts to detect fraudulent activities in the séances and performances of Fox, Home, and Guppy, yet according to Wallace, none of them had ever been caught cheating. This was why Wallace placed so much weight on the investigations that examined their alleged powers.

In the case of Fox, Wallace began by asserting that she was the first prominent medium of the modern spiritualist movement. Her powers were discovered at the age of nine, when she and her sisters allegedly communicated with spirits at their family home in New York state.[89] Since then, Fox's career had blossomed, and she traveled around North America and Europe performing for both private and public audiences. Wallace argued that for twenty-six years "sceptic after sceptic, committee after committee, endeavoured to discover 'the trick;' but if it was a trick this little girl baffled them all."[90] By claiming that Fox had confounded skeptics for nearly three decades, Wallace was attempting to establish the legitimacy of her powers, but he did not stop there, and he included the reports of prominent spirit investigators who had observed and confirmed her mediumistic abilities firsthand.

One of those investigators was the Scottish American social reformer Robert Dale Owen (1801–1877), who wrote two well-known works on spiritualism, *Footfalls on the Boundary of Another World* (1860), and *The Debatable Land between this World and the Next* (1871). Owen had extensive experience investigating spirit phenomena and mediums, and he gained a reputation as a leading expert in the field.[91] Wallace recounted,

> Mr. Owen had many sittings with Miss Fox for the purpose of test; and the precautions he took were extraordinary. He sat with her alone; he frequently changed the room without notice; he examined every article of furniture; he locked the doors and fastened them with strips of paper privately sealed; he held both the hands of the medium. Under these conditions various phenomena occurred, the most remarkable being the illumination of a piece of paper (which he had brought himself, cut of a peculiar size, and privately marked), showing a dark hand writing on the floor. The paper afterwards rose up on to the table with legible writing upon it, containing a promise which was subsequently verified.[92]

Despite Owen's comprehensive experience in detecting fraud, and all of the safeguards in place during the experiments, Fox was still able to produce spirit phenomena. Wallace believed that her powers were legitimate, and Owen was a credible witness.

The Scottish publisher, naturalist, and anonymous author of the *Vestiges of the Natural History of Creation* (1844), Robert Chambers (1802–1871), also had the opportunity to investigate Fox's psychic abilities with Owen during a visit to the United States in 1860.[93] Chambers had a keen interest in spiritualism, possessing a sound knowledge of the literature, and extensive experience attending séances during the 1850s. These factors strengthened the case for him being seen as a trustworthy observer.[94] Together Chambers and Owen conducted some tests on Fox to determine whether her psychic powers were genuine. Wallace described some of the safeguards that Chambers and Owen used during their tests with Fox to ensure that there was little opportunity for deceit. This included the use of gas lighting so that the room was fully visible during the tests, weighing down the séance table with a heavy steelyard so that it was too difficult to move manually without being detected, and insisting on Fox's hands being held over her head and not touching the table. Despite all of these precautions, Fox was allegedly able to produce remarkable psychic feats.[95]

Wallace observed that these control measures were similar to the ones used by the chemist and physicist Michael Faraday (1791–1867) during his investigations into table turning in 1853. After conducting experiments, Faraday proposed that the strange table movements that he witnessed during spiritualist performances were the result of "unconscious muscular action" on the part of séance sitters.[96] Therefore, spirit and psychic phenomena were not real. Wallace disregarded these conclusions because he believed that Faraday was not sufficiently versed in spiritualism, and had not attended enough séances to form a sound conclusion. However, that was not what was important to Wallace. What mattered was that the scientific community had approved of Faraday's experimental method. If it was good enough for him, then it was good enough for any researcher using it. By employing Faraday's methods in their investigation of Fox, Chambers and Owen were strengthening the veracity of their findings. They were also regular attendees of séances, and well versed in the spiritualist literature. Thus, they were far more reliable as witnesses, and their conclusions should be given more weight.[97]

As was the case with Fox, Daniel Dunglas Home's mediumistic pow-ers were allegedly discovered at an early age, and by the middle of the nineteenth century he was one of the most widely known mediums in the world. Originally from Scotland, Home moved to the United States in the late 1830s to live with family in Connecticut. He returned to Britain in 1855, around the same time that prominent skeptics such as Faraday and the physician and zoologist William Benjamin Carpenter (1813–1885) were explicating the causes of spirit phenomena such as table turning through ideomotor responses (or unconscious reflexes).[98] While most Victorian medi-ums were fairly clumsy in their performances, and easily detected as frauds, Home was of a different caliber. His success was largely due to his ability to produce a broad range of extraordinary phenomena during séances in fairly good light. Arguments for unconscious muscular movements alone were insufficient in discrediting his alleged powers.[99]

Spirit investigators regularly examined his amazing mediumistic acts in order to determine whether or not his psychic powers were genuine.[100] One of the notable investigators to study Home's powers was the Scottish physicist and mathematician Sir David Brewster (1781–1868). Wallace included excerpts of Brewster's observations from a sitting with Home, showing examples of unexplainable phenomena. Brewster wrote, "The table actually rose from the ground when no hand was upon it," and he continued by noting that "a small bell was laid down with its mouth upon the carpet, and it actually rang when nothing could have touched it."[101] Although he was a skeptic of spirits and psychic forces, Brewster was unable to detect any fraud during his investigation of Home's mediumistic abilities. He was therefore left baffled by the events that he had witnessed. For Wallace, this sort of evidence was particularly valuable in establishing the legitimacy of spirit phenomena. If a skeptic such as Brewster, who had an extensive background in the physical sciences, was unable to detect any fraud, Home's powers must have been real. To add further credibility to Brewster's account, Wallace noted that the law-yer and politician Lord Henry Peter Brougham (1778–1868) was also present during Brewster's investigation, and he confirmed Brewster's report.[102]

In an effort to assure his readers that Home was an honest and genuine medium, Wallace asserted that Home openly invited investigators to exam-ine his powers, further showing that he had nothing to hide. Wallace re-marked that the lawyer, journalist, and publisher Serjeant Edward William Cox (1809–1879) and the chemist and physicist William Crookes (1832–1819) both investigated Home's powers and detected no deceitfulness.[103] Both Cox and Crookes were reputable Victorian gentlemen, with backgrounds in law and science respectively. For Wallace, they were both credible witnesses. In each case, Cox and Crookes observed what appeared to be authentic spirit and psychic phenomena. For example, referring to Cox's experiments with Home, Wallace wrote, "Serjeant Cox, in his own house, has had a new accordion (purchased by himself that very day) play by itself, in his own hand, while Mr. Home was playing the piano. Mr. Home then took the accordion in his left hand, holding it with the keys downwards while playing the piano with his right hand, 'and it played beautifully in accompaniment to the piano, for at least a quarter of an hour.'"[104]

Agnes Elisabeth Guppy was also discussed in *Miracles and Modern Spiritualism*.[105] Born in London in 1838, Guppy rose through the ranks of Victorian spiritualists during the late 1860s, and became renowned for producing spirit materializations at séances. As Alex Owen recounted, "Guppy perfected the production of spirit 'apports,' usually flowers or pretty gifts, which showered down on the surprised and delighted sitters."[106] It was Wallace who had first discovered Guppy's psychic powers during a séance at his home on November 23, 1866. He soon arranged regular sittings with the young medium every Friday evening for nearly an entire year, so that he could repeat his tests and trace her development.[107] He wrote that on one occasion he secretly tested Guppy's mediumistic abilities by "attaching threads or thin strips of paper beneath the claws [of a table], so that they must be broken if any one attempted to raise the table with their feet—the only available means of doing so."[108] Yet, despite this safeguard, the table still rose without any damage to the strips that Wallace had carefully placed beneath it. For Wallace, this was evidence in favor of genuine spirit phenomena. Wallace's tests using paper strips to determine whether Guppy or one of the other sitters were cheating during the séance was also recorded in his unpublished notebook in the entries for both November 30 and December 7, 1866. However, as his notes indicate, during his second experiment Wallace added further controls, and he wrote, "Beside the tissue paper under the table I had constructed a cylinder of hoops & brown paper within which the table was placed thus keeping dresses and feet away from it; yet it rose up as before. All this with sufficient light to see everything."[109]

Wallace's personal testimony is sprinkled throughout his book, and he included long excerpts of his observations from the many séances that he had attended. In most cases these published descriptions match the information he recorded in his private séance notebook. This was a key component of his strategy for establishing himself as a credible witness of spirit phenomena. It demonstrated not only that he was familiar with the literature on modern spiritualism but that he also possessed direct experience engaging with psychic forces. The emphasis on experiential knowledge was essential for establishing himself as an expert. His examination of spirit phenomena at séances can be treated as a kind of ethnographic study, and much like how scientific explorers verified the trustworthiness of their ethnographic observations in travel narratives, through detailed discussions of their daily activities, Wallace was using a similar method in his writings. This process of outlining the details of an experiment transformed his readers into what Shapin and Schaffer have called "virtual witnesses," allowing them to acquire an almost firsthand knowledge of his investigations. [110] It was therefore a core aspect of Wallace's "visual epistemology." [111]

One example in Wallace's book is a description from his very first séance experience on July 22, 1865. He wrote,

> Sat with my friend, his wife, and two daughters, at a large loo table, by daylight. In about half an hour some faint motions were perceived, and some faint taps heard. They gradually increased; the taps became very distinct, and the table moved considerably, obliging us all to shift our chairs. Then a curious vibratory motion of the table commenced, almost like the shivering of a living animal. I could feel it up to my elbows. These phenomena were variously repeated for two hours. On trying afterwards, we found the table could not be voluntarily moved in the same manner without a great exertion of force, and we could discover no possible way of producing the taps while our hands were upon the table. [112]

There are several important points to emphasize in this passage. To begin, Wallace states that the séance was conducted during the day, making the room fully visible so that any deceitful activity by the medium could feasibly have been seen. Moreover, he was not alone, and there were multiple witnesses observing the same phenomena. In addition, the spirit phenomena were repeated and sustained for a long enough period of time that it was possible for Wallace to carefully observe and take note of what he was witnessing. Finally, all of the objects involved in the séance were inspected immediately afterward to determine whether any trickery was possible. The inclusion of sensory detail, such as the "vibratory motion of the table," added a further layer of authenticity to Wallace's description, allowing for his readers to imagine the sensation on their own bodies.

Wallace continued to experiment with psychic forces, and he conducted repeated tests to determine whether or not he could consistently observe similar phenomena during séances. He stated,

> On other occasions we tried the experiment of each person in succession leaving the table, and found that the phenomena continued the same as before, both taps and the table movement. Once I requested one after another to leave the table; the phenomena continued, but as the number of sitters diminished, with decreasing vigour, and just after the last person had drawn back leaving me alone at the table, there were two dull taps or blows, as with a fist on the pillar or foot of the table, the vibration of which I could feel as well as hear. No one present but myself could have made these and I certainly did not make them.[113]

Once again, we see that Wallace was not alone during the tests, and that there were other witnesses available to confirm his reports. By providing a step-by-step account of his experiments with table rapping, we see a further attempt by Wallace to demonstrate that he was a skilled spirit investigator, which reinforced his claim as being a credible witness.

It is also rather revealing to examine some of the examples of séance phenomena that were not included in Wallace's published accounts from *Miracles and Modern Spiritualism*, as they provide important insights into both his ethnographic "field practice" and commitment to documenting prima facie evidence. In a sense, it helps to reconstruct a "thick description" of Wallace's research methods during his investigations into spirit and psychic phenomena.[114] What is surprising is that some of the most extraordinary displays, which he had observed during the height of his investigations of séances during the late 1860s, were omitted from his book. The reasons for these omissions are unclear, but nevertheless they offer an illuminating glimpse into the world of Victorian spiritualism, and the practices of investigators who studied spirit and psychic manifestations.

For example, we see some of Wallace's early ruminations in his private notebook about spirit and psychic entities and unseen forces. He wrote in his entry from July 22, 1865, that the recent results of his experiments with table turning satisfied him that there was "an unknown power developed from the bodies of a member of persons, placed in connection by all when hands being on a table."[115] His notebook also contains a diagram from these experiments that traces a wavelike pattern that the table allegedly followed as it traveled around the room (fig. 1.1). Wallace believed that these movements were caused by spirit interventions, and he wrote, "The movement of the table was almost always in curves as if turning on one of the claws, so as to give a progressive motion. This would be frequently reversed & sometimes regularly alternate so that the table would travel across the room thus."[116]

The function of this diagram was important for Wallace's visual epistemology, because it allowed him to visually record his impressions from the séance while they were still fresh in his mind, thereby enabling him to recreate the experience for future analysis.[117] The entry also brings to life much of the embodied practice that Wallace used in his field-based approach to studying spiritualism, that is hidden or erased in the published accounts from his book.

Wallace's personal notebook on his spirit investigations is illuminating in other ways. Take, for instance, the series of séances that he participated in at the home of his friend Frederick Lokes Selous (1802–1892), in Regents Park, London. Wallace attended eight séances in total with the Selous family between March 26, 1867, and June 21, 1867. All of them were led by Guppy, and Wallace's descriptions from these events provide far more intimate knowledge of the inner workings of Victorian spiritualist performances than any of his published accounts contain. More importantly, they capture the voices of genuine believers of the spirit hypothesis, who are so often marginalized or censored in the reports of other scientific investigators from the Victorian period. For instance, Wallace's entry from May 29, 1867, recounts how Selous's daughter Florence (1850–1921) was miraculously levitated from her seat and placed on top of the table, much to the delight of the sitters. Wallace wrote, "Miss F. was three times placed on the table, once sitting, twice standing . . . [and] on all occasions silently and instantaneously." To complement these observations, Wallace included sensory information in his notebook, which allowed him to capture his physical impressions from

▶ Figure 1.1. This page from Wallace's personal notebook on spiritualism includes a small diagram that depicts a wavelike pattern that a table allegedly traveled along around a room during a séance that he attended on July 22, 1865. *Source*: Alfred Russel Wallace, Journal, July 22, 1865, p. 42, WCP5223.5749, Papers of Alfred Russel Wallace, Natural History Museum, London.

motion - This would be frequently
reversed & sometimes regularly
alternate so that the table would
travel across the room thus

There is no doubt that the persons present
could move the table as it moved,
with their hands, but the precautions
& experiments proved this could not
be always the case, & we have therefore
no right to suppose it was ever the
case. The raps on the other hand we
could not make it at all. They were small
clear raps as if made with a fairy
hammer. The only approach to them
could be made by tapping with the
finger nail under the table; but as
all our hands were upon the table with
all our eyes constantly open I am certain
they could not have been done by any one present.
The result of these experiments is, that I am
satisfied there is an unknown power developed from
the bodies of a number of persons, placed in connection
by all their hands being on a table. MF.

the séance. He explained, for example, how "no noise or sound" was heard during these levitations, and that despite holding Florence's hand and feeling it the entire time as a control to keep her in her seat, she still ended up on the table—a feat that seemed to genuinely baffle the naturalist. Moreover, all of the attendees believed the levitations to be real products of spirit manipulations, thus further legitimizing the extraordinary phenomena on display through corroborating testimonies.[118]

One of the more impressive manifestations to occur regularly during the séances led by Guppy was the production of spirit apports—especially fresh flowers and ferns—which unexpectedly showered down on the sitters. Spirit apports eventually became the hallmark of Guppy's performances during the 1870s, attracting many Victorians to her commercial séances.[119] Already in the late 1860s, however, these manifestations were considered remarkable, and they feature heavily in Wallace's notebook. For example, the flowers and ferns appearing at the séances Wallace was attending were covered with dew from the outdoors. He recorded in his notebook that "it was too wonderful to be credited yet it was palpably true, for the fresh dewy flowers and ferns could not have been concealed in a hot dry room for more than an hour had any one of the party tried to deceive the rest."[120]

On another occasion in June 1867, Wallace imposed stricter controls during the séance to determine whether any trickery was responsible for producing the spirit apports. To strengthen the veracity of his investigation, he invited the chemist John Hall Gladstone (1827–1902) to act as a credible and scientifically minded corroborating witness to verify his report. Although they sat in "total darkness," during Guppy's performance, "matches and candles [were] ready" to strike and illuminate the room if any transgressions were sensed.[121] Moreover, everyone at the table held hands as a means of limiting the other sitters from conspiring with the medium to produce fake manifestations. Despite these controls, Wallace recorded that "showers of flowers" fell onto the table, including "a fine waterlily." He remarked that the appearance of this particular flower was especially significant because "a lady at the table [who] said her name was Julia . . . had that day been to Covent Garden to try and get a waterlily and had failed." Guppy had no knowledge of this excursion, and yet of all the flowers that could have appeared during the séance, it was a waterlily that manifested. Wallace believed that it was too improbable for this occurrence to be mere coincidence, and he believed that an unseen power was at work.[122]

Reflecting on Wallace's personal testimonies from his notebook, it would seem that he interpreted the majority of the manifestations that he had witnessed at the séances as real products of spirits and psychic forces. However, a skeptic could have maintained that his investigations were conducted under less-than-ideal conditions. More often than not, these séances were done under the cover of total darkness, providing an ideal environment for a clever trickster to manipulate the sitters' perceptions. There were also few controls in place, such as powder to cover the medium's hands, or bindings to limit her movements. The séance rooms were also rarely sealed to prevent others from entering the space once the lights were put out, and many of the sitters attending these séances were unknown to Wallace. Thus, they could easily have been colluding with Guppy to produce fake phenomena. As a source of evidence to legitimize spiritualism, it was flawed. Nevertheless, Wallace's notebook provides fascinating examples of the sorts of extraordinary phenomena that were fundamental to supporting his own belief in the spirit hypothesis, and it effectively captures both his commitment to prima facie evidence and his reliance on field-based ethnographic methods. All of these experiences informed, and were reflected in, his core arguments from *Miracles and Modern Spiritualism*.

Another type of evidence that Wallace placed great emphasis on in his spirit investigations was photographs. During the 1860s, figures such as Huxley and the polymath Francis Galton (1822–1911) were championing the value of photographic evidence for anthropological studies.[123] Wallace's application of this technology in his spirit investigations was following in the latest disciplinary trend. He viewed photographs as one of the most effective forms of evidence for supporting the legitimacy of the spirit hypothesis. Paintings and drawings were seen as subjective visual representations of spirits, because they were influenced by the inherent biases of artists. Photographs by contrast, were different, and according to Wallace they produced verisimilar depictions of spirits that were supposedly mediated through camera technologies.[124] Of course, skeptics argued that photographers manipulated plates all the time by adding fake spirits to them during the exposure process. However, Wallace stated that these forgeries could be caught, and in *Miracles and Modern Spiritualism* he outlined five precautionary measures for limiting the chance of deception. If these precautions were followed, spirit photographs could be given a primal status in the defense of the spirit hypothesis.

The first precaution required spirit investigators to have a sound knowledge

of both the materials required for producing plates and photographic processing. Wallace wrote, "If a person with a knowledge of photography takes
his own glass plates, examines the camera used and all the accessories, and
watches the whole process of taking a picture, then . . . it is a proof that some
[spiritual] object was present."[125] The second precaution considered whether
the likeness of the spirit appearing in a photograph was similar to a deceased
loved one, who "was totally unknown to the photographer."[126] If this was the
case, then it was positive proof that the spirit in the image was real. How else
would the photographer be able to forge the likeness of the spirit? For the
third precaution, Wallace argued that if a spirit appeared in a photograph
that was arranged by a sitter "who chooses his own position, attitude, and
accompaniments," then it was also confirmation that the image was genuine.
For the fourth precaution, Wallace focused on whether there was any evidence
to suggest that the photographic plate had an image superimposed on it. He
wrote, "If a figure appears draped in white, and partly behind the dark body
of the sitter without in the least showing through, it is a proof that the white
figure was there at the same time, because the dark parts of the negative are
transparent, and any white picture in any way superimposed would show
through."[127] Finally, for the fifth precaution, Wallace stated that if a medium
described a spirit before a photograph was taken, and its matching likeness
appeared on a photographic plate afterward, then it was to be taken as proof
that the captured image was the product of genuine spirit phenomena.[128]

Wallace applied all of these precautions to a photographic session he held
with the famous spirit photographer Frederick Hudson (b. 1812), in March of
1874.[129] Guppy was also present on this occasion, acting as a corresponding
witness to the event. Three photographs were taken of Wallace, with each
depicting a different spirit in them, and the details are described in *Miracles and Modern Spiritualism*. According to Wallace, he was present during
the processing of the images, and he chose which poses to hold for the
photographs. The spirit that appeared in the third image was particularly
noteworthy because it was draped in white, and allegedly shared a likeness
to Wallace's mother (see fig. 1.2). Before holding the session with Hudson,
Wallace recounted that he had received "communication by raps to the

▶ **Figure 1.2.** An alleged spirit photograph of Alfred Russel Wallace and his mother
captured by Frederick Hudson on March 14, 1874. *Source*: Georgiana Houghton,
Chronicles of the Photographs of Spiritual Beings (London: E. W. Allen, 1882) between
pages 224–25, plate 6, no. 49.

Died 1869

March 14th. 1874

effect" that his mother would "appear on the plate if she could." Wallace was utterly convinced that the photograph of him and his mother was an authentic example of a spirit manifestation. He wrote, "Even if he [Hudson] had by some means obtained possession of all the photographs ever taken of my mother, they would not have been of the slightest use to him in the manufacture of these pictures. I see no escape from the conclusion that some spiritual being, acquainted with my mother's various aspects during life, produced these recognisable impressions on the plate."[130] Because the photograph seemed to pass all of Wallace's precautionary measures, he viewed it as a credible source of evidence for supporting the spirit hypothesis.

Wallace's aim of developing a "new branch of anthropology" that was devoted to spirit investigations fundamentally relied on his ability to establish the trustworthiness of his evidence. He followed a long tradition in ethnology and anthropology of relying on the accounts of firsthand observers to substantiate his suppositions. He took great care in showing his readers how the sources that he used in his book were highly credible. Most of the accounts that Wallace used were by prominent investigators who possessed both a strong knowledge of the literature on spiritualism, and extensive experience examining spirit phenomena at séances. He supported this information further with his own personal testimony from the séances that he had witnessed firsthand. With the so-called facts that he presented to his readers, Wallace hoped that he made a strong enough case for recognizing investigations of spirit phenomena as a genuine scientific pursuit that showed the reality of spirits and psychic forces.

Changing Anthropological Views and the Rise of Animism

With the publication of Tylor's *Primitive Culture* in 1871, anthropological interest in the study of religion grew. As the next chapter will explore in more detail, spiritualists became the subject of anthropological research, and they were characterized as practicing a "primitive" and "superstitious" form of belief that was indebted to an age when all humans saw the world as being inhabited by spirits—what Tylor called "animism."[131] This was not the response that Wallace had been hoping for when he attempted to establish psychical research as a subfield of anthropology during the middle of the nineteenth century. He subsequently became a vocal critic of Tylor's work.[132] In 1872 he published a scathing review of *Primitive Culture* in the popular review journal the *Academy*. He accused Tylor of lacking theoretical sophistication, and of merely stockpiling ethnographic data without properly interpreting its significance. Wallace framed *Primitive Culture* as a laborious read that researchers were best to avoid.[133]

What really agitated Wallace, though, was that Tylorian anthropology criticized the evidentiary foundation of spirit investigations—an apparent reliance on dubious personal testimonies. Tylorian anthropologists identified many factors that contributed to misperception when interpreting extraordinary phenomena. According to these anthropologists, most reports that professed to observe genuine spirit or psychic phenomena had not been properly vetted by investigators. Had they thoroughly considered the ways in which observers misinterpreted supernormal events, using animistic principles in their analysis, the evidence would most likely lead to a skeptical conclusion. Thus, Wallace's arguments and beliefs were based on a false understanding of the materials. The theory of spiritualism, as articulated by Wallace in *Miracles and Modern Spiritualism*, was incommensurable with Tylor's emerging anthropological paradigm, and debates over the existence of spirits and psychics continued.

Wallace's theories and practices drew on larger disciplinary discussions about how to do reliable ethnological and anthropological research during a period when the boundaries of the discipline were still being negotiated. Through careful consideration of the practices and theories that he used to legitimize his spirit investigations as a genuine scientific pursuit, an important story emerges about how nineteenth-century researchers constructed their truth-claims and became credible witnesses. This is particularly telling because spiritualism deals with nonconventional knowledge that runs counter to what has been deemed "proper" science.

Supernaturalism, as the word implies, violates the accepted laws of nature and undermines, for scientific naturalists, the basic premises of science. Believers in spirit phenomena and psychic forces had to work extra hard to prove the reliability of their evidence.[134] Yet, as Tylorian anthropologists argued, the personal testimonies of observers who had witnessed spirit activities were an insufficient benchmark for laying the foundation of a new science. Although Wallace's efforts ultimately failed to make spirits and psychic forces major areas of specialism in anthropological research, the methods and theories that he developed during this process took root elsewhere. With the formation of the SPR in 1882, spirit investigations gained an organization backing that helped to foster studies into extraordinary phenomena, but the debates over the reality of the spirit hypothesis raged on. Over the next few decades, animism would become one of modern spiritualism's great adversaries in scientific forums, and anthropological figures such as Tylor, James George Frazer (1854–1941), and Edward Clodd its chief opponents.[135]

2

Edward Burnett Tylor

The Skeptic

In the summer of 1871 numerous reports emerged among spiritualists in London that a group of young mediums, including Florence Cook (1856–1904), Frank Herne (b. 1849), and Charles Williams (b. 1848), were able to produce fully formed, physical manifestations of the spirits known as John and Katie King—the alleged spiritual identities of the seventeenth-century Welsh privateer Sir Henry Owen Morgan (1635–1688) and his daughter, Annie Owen Morgan.[1] Herne and Williams had acted as mentors to Cook, ushering the teenager into the world of Victorian spiritualism, even holding several joint sittings together at Herne and Williams's séance room at 61 Lamb's Conduit Street, near Russell Square, in London. The spectacle surrounding these performances drew a lot of attention; Cook in particular was revered as a powerful spirit materialist, and was soon ranked among the elite mediums in Britain.[2] It was not long before other mediums claimed to be producing various sorts of similar spirit phenomena at séances. These reports regularly appeared in spiritualist publications, and formed the foundation of many studies that aimed to prove the existence of spirits and psychic forces.[3] Amazed by what he had read in these published accounts, the ethnologist-turned-anthropologist Edward Burnett Tylor decided that he had to see for himself whether or not there was any truth to these unbelievable claims. As an ardent secularist, who argued that a belief in spirits was a cultural survival of primitive thought, Tylor doubted the veracity of these reports.[4]

Up until the early 1870s Tylor had limited direct experience with spiritualism. He had attended a spiritualist performance in the 1860s, but it was a complete sham, and it had only heightened his skepticism of the movement. In a letter to Alfred Russel Wallace from November 26, 1866, Tylor discussed how he had attended a séance at Hanover Square in London, where the so-called medium was exposed as a fraud. It was discovered, according to Tylor, that the purported spirit phenomena witnessed during the séance were forged with the help of "paid subjects," who were planted in the audience. This left a lasting impression on Tylor, leading him to conclude that all spirit phenomena were faked, and every séance a performance in the art of deception.[5] However, the continued appearance of spiritualism in books and periodicals could not be ignored, and slowly Tylor's interest in spiritualism grew again—albeit as curious critic and skeptic, not as a proponent of the movement.

The extraordinary nature of most accounts of spirit phenomena brought into question the credibility of the witnesses who had professed to observe them. As Tylor remarked in his letter to Wallace, "So far as I know anything of . . . believers in spiritualism I cannot say I like them as observers in a field of enquiry particularly haunted by professional impostors."[6] If these secondhand accounts were to be taken seriously as reliable evidence for spirit investigations, the trustworthiness of the observers needed to be established. Like most anthropologists in the nineteenth century, Tylor was highly skilled in analyzing secondhand accounts to determine their reliability. This process was founded on a number of factors, including whether the background and training of observers who recorded such accounts gave them sufficient expertise in assessing the phenomena they witnessed, and if the observations they produced matched other sources that were deemed credible.

To legitimize the reliability of an observer's testimony, anthropologists such as Tylor had to compare and contrast different records, and base their conclusions on the most commonly observed phenomena in the various texts. It was a sort of "collective empiricism."[7] If a testimony contained discrepancies or inconsistencies with other accounts, the anomalous one was deemed untrustworthy. These were rough waters to sail, and a source that was considered trustworthy by one person could be seen as spurious by another. In these instances, researchers could add greater authority to their views if they supported them by combining a sound knowledge of the literature and testimony of other witnesses with their own firsthand experience investigating spirit phenomena.[8] Tylor's approach to assessing the reliability of spiritualist accounts shared much with the methods Wallace articulated in *Miracles and Modern Spiritualism*. However, there was an important distinction: Wallace was driven by his firm belief in the legitimacy of modern spiritualism, and Tylor was skeptical of it. Thus, while their approach may have been similar, their findings were markedly different.

Tylor wanted to test the veracity of his theory of animism, and in November 1872 he traveled to London so that he could undertake an ethnographic inquiry into the modern spiritualist movement. During this period of investigation, he kept a notebook, which contains his observations of the séances and other spiritualist performances that he attended during his visit to the capital. In the opening paragraph of his notebook he wrote, "I went up to London to look into the alleged manifestations. My previous connexion with the subject had been mostly by way of tracing its ethnology, & I had commented somewhat severely on the absurdities shown by examining the published evidence."[9] *Ethnology*, as Tylor is using it here, meant the historical, armchair-based study of human culture. In other words, he had been tracing the rise and growth of modern spiritualism since its emergence in the late 1840s through printed sources. He remained utterly dissatisfied with the information available, however, because so much of it contained unbelievable "absurdities." Tylor wanted to challenge these extraordinary claims by seeing for himself whether there was any truth to them. This meant that he had to undertake a sustained period of ethnographic study collecting prima facie evidence, which gave far more weight to any study than a hundred secondhand accounts purporting something to the contrary. If Tylor was going to challenge the spirit hypothesis, and position himself as a credible witness and authority on the subject, he had to enter the world of the Victorian spiritualists.[10]

Tylor's interest in spiritualism was intricately tied to his research on the anthropology of religion. For much of the 1860s and 1870s, Tylor had been developing one of his most important and long-lasting contributions to the discipline—his theory of animism.[11] His core argument was that all religions evolved from a rudimentary belief in spirits animating the world. By identifying the laws that governed the development of religion, Tylor attempted to plot all forms of worship onto an evolutionary scale that showed how religious beliefs transformed from basic understandings of the world being animated by spirits, to complex religious systems such as Christianity. Central to his purpose was an attempt to naturalize all religions and explain their ontologies using scientific theories. He was not trying to reconcile science and religion, but to bring religion under the domain of scientific understanding. Spiritualism entered this discussion because Tylor saw it as a survival of primitive belief. In his view, there was little difference between the spirit control of a medium in London, and the spirit possession of a shaman in Africa. Challenging the legitimacy of the spirit hypothesis was an important test for demonstrating the validity of the theory of animism, because if spirit phenomena were proven to exist, Tylor's whole argument would unravel.

Although many scholars have discussed aspects of Tylor's anthropological research program, he remains a surprisingly underexplored figure.[12] Despite his major contributions to the disciplinary development of British anthropology, he has been the subject of only one major biography, which was written by his former colleague and friend, Robert Ranulph Marett (1866–1943), and published in 1936.[13] Even sparser in the historiography is a sustained examination of Tylor's investigation of spiritualism. George Stocking published a transcription of Tylor's notebook in 1971 with a short introduction, but no one has undertaken a detailed study of it.[14] Tylor evidently felt that spiritualism's regular appearance in published works meant that it had to be taken seriously, so much so that he left the confines of his armchair to investigate the matter firsthand. A more rigorous examination of Tylor's investigation into the modern spiritualist movement, which marks an important period in his career, is clearly warranted.

Tylor was not simply an armchair theorist, as he is normally positioned, but a skilled ethnographer working in the field. His early ethnographic experiences traveling through the Americas during the 1850s were crucial to his establishment as a credible observer of human culture. These experiences laid the foundation for his later authority as a world-leading anthropologist. His theory of animism, developed during the 1860s, was formalized with the publication of his most famous book, *Primitive Culture*, in 1871. It was there that Tylor produced his most mature and detailed understanding of animism: the book's entire second volume was devoted to it. Throughout this period of theoretical development, Tylor regularly compared the modern spiritualist movement to various forms of extra-European primitive belief, including shamanism. While his analysis was based on accounts published in both the periodical press and major spiritualist works, it was not supported by any prima facie evidence from his own experiences at spiritualist performances. It was all collected through secondhand sources. That all changed, however, in November of 1872, when Tylor traveled to London in order to observe, and engage with firsthand, the phenomena produced by mediums at séances. During his visit, Tylor attempted to rationalize and explain with natural causes, the phenomena that he was witnessing during spiritualist performances. His "visual epistemology" was about making sense of these so-called supernatural feats through ethnographic inquiry.[15]

Fundamentally, Tylor's trip to London was all about establishing himself as a credible observer of spiritualism, who could dispel the legitimacy of the spirit hypothesis once and for all. Yet things became far more complicated. Tylor was unable to explain easily how all of the various phenomena at the performances and séances were produced, and his views on spiritualism changed considerably as a result. He remained a skeptic, but not an unfaltering one. Tylor admitted that some of the displays that he had observed during his trip to London sensitized him to the spiritualist appeal, insofar as he was willing to consider future evidence that could prove the reality of the spirit hypothesis. As he became more immersed in the culture of spiritualism, he found himself becoming increasingly receptive to the movement, even befriending some of its most significant proponents. He did not become a believer like Wallace, but he did develop a more nuanced appreciation of the spiritualist movement. Questions of whether spiritualism should be taken seriously, therefore, remained unresolved.

Edward Burnett Tylor and the Making of a Skilled Ethnographer

From a young age Tylor was deeply sensitive with respect to different cultures and races. He was born into a "middling-sort" Quaker family, and received most of his education at Grove House School in Tottenham, which was owned and operated by the Society of Friends.[16] Most Quakers were committed to religious and racial tolerance, and they were inclined to be relativistic and sympathetic toward people of different beliefs and racial backgrounds. Many leading ethnological figures in the first half of the nineteenth century were Quakers, including the physicians James Cowles

Prichard and Thomas Hodgkin (1798–1866). Quaker religious beliefs influenced several of the discipline's central tenets. For example, the Quaker doctrine of the "inner light" put considerable emphasis on the idea of all humans being equal, and this aligned easily with monogenesis—a theory that postulated a common human ancestral origin.[17] By highlighting the physical and social similarities between all races, monogenetic theories supplied Quaker ethnologists with scientifically grounded arguments against the exploitation of extra-Europeans, and generated support for cultural and racial equality through moral, physical, and intellectual improvement.[18]

Tylor's immersion in this social milieu, where issues of cultural and racial parity were at the forefront of communal discussions, shaped his later work in anthropology. Throughout his long career he was a staunch monogenist, and, in his writings he was a proponent of human unity.[19] When it came to investigating the belief systems of different cultures around the world, he was careful to maintain an objective tone, insisting that it was important to understand the historical and contemporary circumstances that created various religions.[20] Spiritualism presented no inherent conflict with Quakerism per se, and members of the Society of Friends would have been tolerant toward people supporting the spirit hypothesis even if they themselves did not endorse it. Thus, it is hardly surprising that Tylor wanted to give a fair hearing to spiritualism. He viewed it as another example of the theistic framework that underlined all religious paradigms—a belief in spirits animating the world. For Tylor, all cultural communities transitioned from a state of religiosity to scientism, and understanding different belief systems was an important part of mapping the process of secularization. As with any belief system, spiritualism deserved attention.[21]

It was not until the 1850s, however, that Tylor's interest in ethnological and anthropological topics began to blossom. After the death of his parents in 1852, Tylor began working in his family's brass foundry. Within a few years he developed tuberculosis, which forced him to change careers. His older brother, Alfred (1824–1884), encouraged him to visit North America to clear his lungs, and taking his brother's advice, the younger Tylor set out on a two-year trip in 1856.[22] His travel experiences at this time represent a formative period in his ethnological and anthropological training, because it was during this trip that Tylor engaged for the first time with extra-European cultures, and developed his skills as a proficient ethnographer. He became a skilled fieldworker and learned to value direct observation. All this would later prove to be essential to his self-positioning as a trustworthy observer of spiritualist performances. In Tylor's view, the séances that he attended in London were no different from the rituals that he had witnessed in Latin America. They could be studied analogously.

In the introduction to his travel narrative, *Anahuac* (1861), Tylor wrote that he spent "the best part of a year" traveling down the Mississippi River, observing the North American Indigenous peoples and enslaved Africans he encountered along the way. He also lived for a short time on a sugar plantation in Louisiana, before deciding to visit Cuba for a new adventure.[23] While in Havana he met the ethnologist, archaeologist, and banker Henry Christy (1810–1865). This chance encounter proved to be an important one, as Tylor's ethnographic talents flourished thanks to his friendship with Christy. The two men had much in common and quickly established a strong rapport. Both were Quakers, had grown up in London, and came from middling-sort families. Moreover, they also had similar educational backgrounds, having studied at schools operated by the Society of Friends. Christy became Tylor's mentor over the following ten years, until his death in 1865. He taught Tylor how to observe ethnological subjects in situ, as well as to explain the major tenets of Prichardian monogenesis.[24] Soon after their first meeting, Christy invited Tylor to accompany him on a four-month horseback journey through Mexico.[25]

Tylor's travels through Mexico had a significant impact on his later ethnological and anthropological writings. By traveling to Mexico and seeing Indigenous peoples for himself, he could claim an authoritative understanding of ethnological subjects, as well as collect substantive data upon which to base his research claims. Under the guidance of his mentor Christy, Tylor had a kind of intensive practical training course in ethnography. He came to realize that one could not rely solely on secondary accounts, whose authors often misrepresented or misunderstood aspects of extra-European cultures, and had no knowledge of different races. As a travel writer, Tylor also became familiar with the practice of informed writing that was necessary for verifying his observations. He understood the significance of mapping his own reports onto those recorded by previous travelers, showing that they were not only consistent with previous accounts, but also moved beyond them with new information.[26] When it was possible, he offered correctives to reports containing misinformation. All of this experience would serve him well in establishing himself as a credible witness of spiritualist activities.

When Tylor returned to Britain in the late 1850s, his status quickly rose within the ethnological community. By the early 1860s he was already a leading figure at the Ethnological Society of London (ESL), with close ties to both the younger generation of scientific naturalists who were becoming increasingly active in ethnological circles and the older guard of Prichardian monogenists, who were on the decline due to both their advanced age and shifting conceptions of science in the middle of the nineteenth century.[27] Tylor even had connections to the Anthropological Society of London (ASL), serving as the society's foreign secretary from 1863 to 1864 until—according to George Stocking—he was enraged by James Hunt's "pugnacious racism," which "offended his humanitarian Quaker beliefs."[28] Tylor did not, however, join Thomas Henry Huxley's attack on the ASL in the 1860s, and because he distanced himself from those debates, he was not a target of Hunt's staunch criticisms of the ethnological community. This might explain why, during the aftermath of the anthropological schism in the early 1870s, Tylor was able to continue to build upon his first-rate reputation; his name was not directly associated with either camp.[29]

Thanks to his family's successful business, Tylor could devote himself entirely to the study of ethnological and anthropological topics. He was thus able to produce many important works throughout the second half of the nineteenth century, including *Researches into the Early History of Mankind* (1865), *Primitive Culture*, and his textbook, *Anthropology* (1881). Despite his financial independence, he ultimately opted for a life in academia. In 1884 he was appointed keeper of the Natural History Museum and reader in anthropology at Oxford. He became a full professor in 1896, remaining at Oxford until his retirement in 1909.[30]

Notwithstanding all of these remarkable accomplishments and contributions to the research field, it was *Primitive Culture* that cemented his status as a world-renowned anthropologist. The book represented the maturation of his cultural theories. Although it is not explicitly stated in his notebook on spiritualism, it is likely that Tylor felt it necessary to investigate spirit phenomena so that he could uphold his theory of animism. If the spirit hypothesis was proven to be false, spiritualism was simply a survival of primitive thought and nothing more. If it were proven to be real, however, animism was a false doctrine, and the foundation of Tylor's writings would collapse. The verified existence of spiritual entities would mean that ancient animistic beliefs were possibly legitimate observances, and not superstitious misperception. With this in mind, Tylor had to show conclusively that the phenomena witnessed at séances, and other spiritualist performances, were not produced by spirit or psychic agencies. As we will see in the next section, Tylor was already challenging the validity of the spirit hypothesis in the 1860s when he was still developing his ideas on the evolution of religion. His ethnographic investigation in 1872 was a continuation and strengthening of this long-standing critique.[31]

The Theory of Animism and Tylor's Early Critique of Modern Spiritualism

After returning from the Americas in the late 1850s, Tylor's research program shifted primarily to the armchair, and he became more theoretically sophisticated in his writings. Gradually, through extensive reading, his familiarity and knowledge of ethnological and anthropological topics broadened, and he based much of his ideas on the testimonies of other writers.[32] As he refined his views on the evolution of religion during the 1860s, his own religious perspective changed, and he was increasingly drawn to rationalism and positivism. Tylor eventually gravitated away from Quakerism, and officially left the Society of Friends in the summer of 1864.[33] Although he never openly called himself an agnostic, the theories that he articulated in *Primitive Culture*, as well as his association with other prominent agnostics such as Huxley, suggest that he adopted aspects of this ideological movement into his research program and worldview.[34] By the late 1860s Tylor's work had become increasingly more secular in tone, and he wanted to produce a naturalistic explanation for religious belief.[35] Spiritualism became entangled in these discussions because Tylor viewed it as a kind of religion.[36]

Before the publication of *Primitive Culture* in 1871, Tylor produced several earlier versions of his theory of animism. These appeared in a series of lectures and articles throughout the 1860s.[37] For example, in 1869 Tylor presented a lecture at the Royal Institution (RI) titled "On the Survival of Savage Thought in Modern Civilization." Central to its purpose was an argument that positioned modern spiritualism as a remnant of primitive belief. So-called acts of supernaturalism were witnessed in all cultures, but that did not mean that these phenomena were genuine. It merely suggested that when no mechanistic or naturalistic cause could be identified, there was a tendency for all human groups to interpret these extraordinary phenomena through irrational (or primitive) explanations. The spirit manifestations allegedly being produced at séances were in essence part of a legacy rooted in primitive religious thought, and this so-called primitive reasoning could be traced historically in Europe.[38] After all, analogous spiritual phenomena were often recorded in medieval religious accounts. These occurrences were usually framed as types of miracles.[39]

Levitation was a typical example of a supernatural phenomenon regularly witnessed around the world and reported in historical accounts. Much of the evidence to support these claims was questionable, and based on dubious personal testimonies that could not be fully verified.[40] Tylor wrote, "One of the most celebrated of modern spiritual manifestations is the feat of rising in the air. This, if not savage, has a long and curious ethnographic history."[41] Cases of levitation could be found among Buddhists in Asia, "where every saint who has attained to 'riddhi,' or perfection, is able to rise in the air."[42] Likewise, it could be found in the chronicles of early Christian writers, and Tylor stated that "*The Lives of the Saints* swarm with it. St Dominic, St Dunstan, St Philip Neri, St Ignatius Layola, are among the list of saints who not only metaphorically 'rose above the earth,' but were thought, particularly by biographers a long while after they were dead, to have literally hung suspended in the air in life."[43] Tylor contended that all of these so-called acts were doubtful because there was no way of confirming whether they truly happened without more concrete data to substantiate the claims. With regard to Tylor's nascent theory of animism, however, that was not the point. What mattered was that this belief in the reality of levitation through supernatural forces was repeatedly recorded across all cultures and ages. It survived in modern European society through spiritualism, and according to Tylor, "some remnant" of the belief that people could levitate through supernatural forces "has now descended on Mr. Home."[44]

One of the most celebrated psychic acts ever to be performed by the medium Daniel Dunglas Home occurred on December 13, 1868, at Ashley House in Victoria, London. In front of several eyewitnesses, including Alexander William Crawford Lindsay, Twenty-Fifth Earl of Crawford (1812–1880); Edwin Richard Wyndham-Quin, Third Earl of Dunraven and Mount-Earl (1812–1871); and Captain Charles Griffith Wynne (1815–1874), Home allegedly levitated from the ground, floated out of a window on the third story, where the séance was taking place, and re-entered the building through a second window in an adjacent room. According to reports this was all done in broad daylight, much to the shock and delight of the séance attendees.[45] Yet several inconsistencies in the story started to emerge as it was retold in the press and in other larger spiritualist works. First, some of the witnesses stated that the séance did not occur during the day, but in the evening with good gas lighting. Others argued that the only light available was from the glow of the moon. Second, there were also discrepancies over which room Home was standing in when he initially levitated from the ground. Some reports argued that Home started in the adjacent room, where he was not fully visible. Witnesses could hear a window being opened before Home suddenly appeared in the main room. Third, there were disputes over whether or not anyone had actually seen Home rise from the ground, or if it was merely assumed so because of his sudden emergence through the window.[46] The simplest explanation for how it was possible for Home to accomplish this feat was by attributing it to supernatural causes. However, with so many contradictions in the narrative, skeptics such as Tylor doubted the legitimacy of the affair. Relying on a spiritualistic explanation of the event was, in Tylorian terms, the product of primitive thinking that would always interpret extraordinary acts as deriving from spiritual forces.[47]

Other analogous forms of spirit manifestations were present across cultures and historical periods. Spirit communication was a common occurrence in many travel narratives, and Tylor argued that "the modern spiritualist is absolutely identical with that of the savage."[48] The only difference was that the sorts of spirits who were supposedly visiting séance rooms in cities such as London, Paris, New York, or Philadelphia were literate, compared to those appearing in so-called savage communities. The spirits in European and American parlor rooms could therefore write messages to mediums and their sitters, and rapping or knocking messages to séance attendees was a distinctly Western phenomenon. Tylor wrote that "savages do not seem to have selected a special class of knocking spirits."[49] Yet, as Tylor contended, these "knocking spirits" appear throughout European history, often in "civilized folklore"—that is, the traditional customs, popular myths, and stories of European peasantry, which Tylor viewed as representing a more primitive state of civilized society. This was exactly the sort of cultural environment where supernatural explanations could flourish.[50]

Despite some of the challenges that critics could level at Tylor's theory, the evidence, no matter how you looked at it, put forward a strong case for conceiving of spiritualism as a cultural survival. Tylor concluded his analysis of the modern spiritualist movement by stating that "even supposing the alleged spiritualistic facts to be all true, and the spiritualistic interpretation of them sound, this does not alter the argument. It would prove that savages were wise, and that we civilized fools have degenerated from their superior knowledge. But it would remain true that modern spiritualism is a survival and revival of savage thought, which the general tendency of civilization and science has been to discard. This is the case of spiritualism as seen from an ethnographic point of view."[51] For Tylor, his theory of animism had been weighed carefully, and he still believed that it maintained its veracity. Nevertheless, as he continued to revise and strengthen his theoretical model, his study of spiritualism, which formed part of his larger analysis of religious belief systems, expanded further with even more global examples. These discussions were included in *Primitive Culture*.

As the 1860s progressed Tylor was gradually drawn toward scientific naturalism, and by the time he published *Primitive Culture* in 1871, this ideological framework was deeply rooted in his theories.[52] Tylor wanted to explain all aspects of human cultural evolution through natural causes. He wrote, "The world at large is scarcely prepared to accept the general study of human life as a branch of natural science. . . . To many educated minds there seems something presumptuous and repulsive in the view that the history of mankind is part and parcel of the history of nature, that our thoughts, wills, and actions accord with laws as definite as those which govern the motion of waves, the combination of acids and bases, and the growth of plants and animals."[53] Tylor's insistence on establishing anthropological research along naturalistic lines was part of his program to secularize the research field and push religious explanations of the world to the margins of the discipline. By reconfiguring anthropological discourse in this manner, belief systems such as modern spiritualism would become topics of anthropology's gaze. The focus would not be on the spirit phenomena that were purportedly channeled through mediums but instead on the reasons why spiritualists believed these acts were genuine examples of psychic forces. It was not about the phenomena itself, but the human interpretation of what was being witnessed. The manifestation was generated by a person and therefore had natural origins.[54]

According to Tylor, all cultural paradigms were essentially the legacy of previous generations readapted into new circumstances. The task for the ethnographer was to understand how human culture transformed through each successive stage, by looking at how different technologies, habits, customs, and belief systems changed over time and appropriated new meanings. Spiritualism, like any belief, could be tracked in this way, although through different routes. It could be studied by tracing changing understandings of spiritual entities across different cultures and periods. It could also examine types of spirit phenomena such as levitation or table rapping to see if it regularly appeared in historical and contemporary records. If there was insufficient data available on a specific group—because of a lack of sources—analogies had to be drawn. For instance, under this model, so-called savage races were seen as analogous to prehistoric Europeans. The two groups could explain one another and even suggest how savage cultures might progress based on European history. The comparative method was thus a core feature of Tylor's cultural theory.[55]

The next step for Tylor in accepting his evolutionary theory of religion was acknowledging that there was much to be gained by studying "savage" beliefs. He wrote, "Far from its beliefs and practices being a rubbish-heap of miscellaneous folly, they are consistent and logical in so high a degree as to begin, as soon as even roughly classified, to display the principles of their formation and development."[56] All humans were naturally predisposed to animistic thinking, and the most rudimentary forms of religion were organized along cogent principles. The ethnographer could follow their development as they transitioned into ideologies more akin to modern European belief systems. Moreover, understanding these transformations would enable ethnographers to determine the roots of religion, and explain why humans accepted the existence of God in the first place. Tylor warned his readers that any study of religion had to contain a fairly broad definition of what constituted a belief system. If it were too specific, it would eliminate many forms of belief that were essential to understanding the underlying reasons for religiosity. His baseline for tracing religious development was his theory of animism—the idea that at the core of all religions there is the notion that spirits (or supernatural forces) animated the world. If researchers were to accept this concept as the basis of their investigations it would provide them with a much more encompassing framework for studying all forms of religion. Spiritualism, therefore, fit easily into Tylor's classification system of religious beliefs; hence its inclusion in his study.[57]

In chapter 11 of *Primitive Culture*, Tylor examined various conceptions of souls and spirits over time and geographic space through a global perspective. Much of this analysis had broader implications for Victorian understandings of modern spiritualism. Many analogies were drawn between modern interpretations of these concepts, and primitive ones. For example, Wallace's version of the spirit hypothesis from *Miracles and Modern Spiritualism* had many similarities to the so-called primitive interpretations of souls and spirits in Tylor's *Primitive Culture*. For Wallace, the spirit was "the essential part of all sensitive beings, whose bodies form but the machinery and instruments by means of which they perceive and act upon other beings and on matter. It is 'spirit' that alone feels, and perceives, and thinks—that acquires knowledge, and reasons and aspires. . . . It is the 'spirit' of man that is man."[58] Yet, as Tylor contended, primitive cultures had a similar view and "the conception of a personal soul or spirit among the lower races . . . is a thin unsubstantial human image, in its nature a sort of vapour, film, or shadow; the cause of life and thought in the individual it animates; independently possessing the personal consciousness and volition of its corporeal owner, past or present."[59] Other comparisons can be drawn. Wallace argued that spirits could "traverse the earth to any distance, communicating with persons in remote countries," and Tylor stated that spirits in primitive cultures could allegedly "separate from the body of which it bears the likeness."[60] With so many traits in common, it was clear to Tylor that modern spiritualism was perpetuating a rudimentary belief grounded in primitive thought. It was therefore a cultural survival and revival from the earliest ages of humankind.[61]

Tylor's discussion of primitive conceptions of souls and spirits, and their links to contemporary belief systems such as modern spiritualism, continues over several pages in his book. It was anchored in the reports of other writers who had collected the information either by traveling to remote regions of the world and engaging firsthand with extra-European cultures, or by participating in spiritualist performances led by mediums in parlor rooms across Europe and North America. As an argument against the validity of the spirit hypothesis it was compelling, but limited. Fundamentally, most spiritualists such as Wallace would have argued that Tylor lacked sufficient direct experience observing spirit phenomena in situ. Without this experience, Tylor's theories would do little to convince spiritualists that their beliefs were indebted to primitive thought. After all, Wallace argued that the truth of the spirit hypothesis could be proved only "by direct observation and experiment."[62] If Tylor were to defeat the spirit hypothesis with naturalistic (or rational) explanations for the causes that justified human belief in spiritualism, it would be done only by entering the séance room and collecting his own prima facie evidence. The theories and methods outlined in *Primitive Culture*, and earlier iterations of it, would underpin his views going in. This was the ultimate test for the theory of animism. In some respects, it marked a foundational moment in the configuration of the subfield of anthropology of religion, because it meant that Tylor was now applying his theories to ethnographic fieldwork. It was no longer the ruminations of an armchair theorist but the observations of a credible witness and fieldworker. It was not so different from the investigatory approach used by Wallace to support his findings in *Miracles in Modern Spiritualism*, save for one major difference: Tylor was not a proponent of the spirit hypothesis but a skeptic who was cynical of the spiritualist movement's core beliefs and practices.

A Skeptic among the Faithful

The application of a "thick description" is particularly useful when reading Tylor's notebook on spiritualism.[63] By closely scrutinizing the details of Tylor's experiences investigating mediums firsthand, we come to appreciate fully how he constructed the robustness of his observations, and positioned himself as a credible witness of spirit and psychic phenomena. It is equally important to position Tylor's findings as those of a skeptic. Although he is more generous compared to other skeptical investigators in giving some agency to spiritualists, his inherent biases are still visible, and they shaped the representations in his notebook. It is therefore essential to locate the politics of writing embedded in his fieldnotes.[64] Only by examining the premises and assumptions underlying Tylor's personal writings, and contextualizing them within their historical context, do we come to recognize his "visual epistemology," and the processes by which he created his ethnographic knowledge.[65] While this may at first glance seem hyperfocused, it does reconstruct an important example of how anthropology was practiced during the late Victorian period. Tylor was, after all, one of the leading figures in anthropology at the time, and he occupied an important role in developing the modern discipline's foundation.[66]

Even before Tylor traveled to London in November of 1872 to probe into the modern spiritualist movement, he had already begun investigating mediums firsthand by participating in a set of experiments led by William Crookes. Since April of 1870 Crookes, with the help of his brother, Walter (1837–1874), had been experimenting on Home to determine whether his mediumistic powers were genuine. Crookes had generated all sorts of numerical data that purported to show decisively that Home was able to produce some sort of unseen "psychic force."[67] The weight of his deliberations was grounded in his commitment to quantifiable evidence that was replicable through experimentation. As a result of this supposedly positive data, Crookes became a believer in the genuineness of mediumship.[68] For many scientific figures during the second half of the nineteenth century, quantifiable data was deemed more reliable than other sources of evidence, such witness testimony, because it gave the appearance of being more objective and freer of personal biases.[69] Crookes therefore strategically used this numerical evidence to reinforce his claim of being a credible witness of spirit phenomena. His expertise as a scientist, who was accustomed to observing and measuring physical phenomena that at times was nearly undetectable, also strengthened his credentials.[70] Despite being a skeptic of spiritualism, Tylor likely viewed Crookes as a reliable spirit investigator with whom he could develop his knowledge of the subject. Their shared network of friends and colleagues probably further convinced Tylor to attend Crookes's experimental sessions. If anyone could persuade Tylor of the reality of spirits, it would be an elite scientific practitioner such as Crookes.

In the spring of 1872 Tylor joined the Crookes brothers during one of their sessions with Home, and recorded these experiences in his notebook on spiritualism. Tylor stated that they began their experiments by sitting around a table, and waiting for spirit communications. Eventually there was some rapping, but the group was overall unimpressed by the occurrences. The investigation continued, though, and other extraordinary phenomena were observed. Tylor wrote that an "accordion under the table held wrong end up by Home played a few notes as if by a player accustomed to the instrument [were] touching them." Next, a table was "tilted & made light and heavy as anyone wished and asked."[71] At one point the table seemingly rose from the ground, and Tylor recounted that he and an unnamed Australian physician, who was also participating in the experiments, attempted to locate the forces that were controlling the table. They even went under the table during the various effects to inspect it, but they were left confounded. After witnessing these various so-called spirit manifestations, Tylor stated that he "failed to make out how either [the] raps, table-levitation, or accordion playing were produced."[72]

Reflecting on his experiences at William and Walter Crookes's experimental session, Tylor believed that Home was an ingenious trickster with an uncanny ability to charm and manipulate people into trusting him. He wrote, "Home is clever, speaks several languages [and is] said to sing and play [music] with much feeling." All of these qualities contributed to the sort of dignified persona Home was trying to self-fashion. He framed himself as a respectable socialite, savvy in the ways of European high society. He even claimed to be the illegitimate grandson of the Tenth Earl of Home. His status was further elevated because he never accepted monetary payment for his performances. By constructing this self-image, Home was able to build a sterling reputation as an exceptionally powerful medium who was able to produce genuine spirit phenomena.[73] Nevertheless, Tylor remained unconvinced by the displays he witnessed. He stated that his "distrust was excited by Home's cleverness, and the way in which he could get pretty women . . . to dance with joy sitting down to his performance, paw him about, [and] call him Dan, etc., which all his intimates have to do."[74]

Despite his suspicions regarding Home's character and overt intimacy with female sitters, the real issue was how the medium was able to produce his spirit manifestations. Unfortunately, Tylor was unable to solve this mystery, and was left stumped by the events that he had witnessed. He argued that some sort of manipulation was at work, and asserted that those who believed in the legitimacy of Home's powers were clearly enchanted by the medium's charisma. The uncertain conclusions resulting from Tylor's investigation of Home led to a further correspondence with William Crookes on the topic of psychic investigations more generally. Tylor's relationship with Crookes also served as an entry point to the next stage of his ethnographic research on the modern spiritualist movement. His first port of call when commencing his more focused study of mediums was at a séance held at William Crookes's house on November 4, 1872.

On this occasion Crookes had invited another medium, named Jennie Holmes, to lead the séance. The spirit investigator Serjeant Edward William Cox, who regularly collaborated with Crookes in his examinations of mediums, was also present.[75] Originally from Philadelphia, Holmes, accompanied by her husband Nelson, traveled to Britain amid a wave of American mediums seeking to escape the horrors of the Civil War between

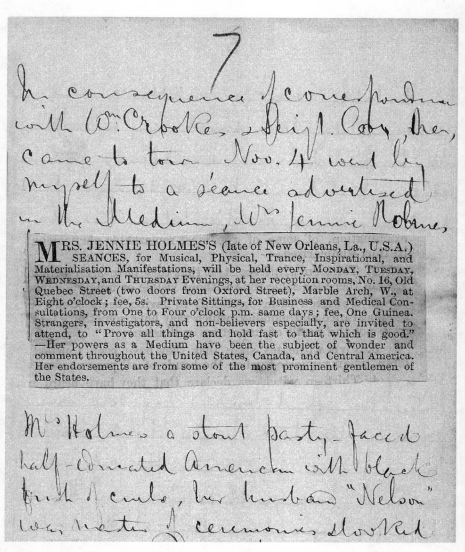

7

In consequence of correspondence with Wm Crookes, Miss Cook, then, came to town Nov. 4 went by myself to a séance advertised in the Medium, Mrs Jennie Holmes.

MRS. JENNIE HOLMES'S (late of New Orleans, La., U.S.A.) SEANCES, for Musical, Physical, Trance, Inspirational, and Materialisation Manifestations, will be held every MONDAY, TUESDAY, WEDNESDAY, and THURSDAY Evenings, at her reception rooms, No. 16, Old Quebec Street (two doors from Oxford Street), Marble Arch, W., at Eight o'clock; fee, 5s. Private Sittings, for Business and Medical Consultations, from One to Four o'clock p.m. same days; fee, One Guinea. Strangers, investigators, and non-believers especially, are invited to attend, to "Prove all things and hold fast to that which is good." —Her powers as a Medium have been the subject of wonder and comment throughout the United States, Canada, and Central America. Her endorsements are from some of the most prominent gentlemen of the States.

Mrs Holmes a stout pasty-faced half-educated American with black frisk of curls, her husband "Nelson" was master of ceremonies slovenly

Figure 2.1. An advertisement for Jennie Holmes' spiritual performances from an unreferenced periodical pasted into E. B. Tylor's notebook on spiritualism from 1872. *Source:* Edward Burnett Tylor, Notebook on Spiritualism, item 12, 7, box 3, Pitt Rivers Museum Manuscript Collection, University of Oxford. This advertisement likely came from the following issue of the *Medium and Daybreak* from November 1, 1872, which was just prior to Tylor's arrival in London to study the spiritualist movement. The typescript is identical to the version pasted in Tylor's notebook: "Mrs. Jennie Holmes's Séances," *Medium and Daybreak* 3 (1872): 436.

1861 and 1865. Many of them stayed during the aftermath of the conflict as the American economy slowly recovered. By the 1860s the spiritualist movement was growing steadily in Britain, and mediums in cities such as London could attract multiple audiences per day.[76] Holmes reaped the benefits of this growing British interest in spiritualism, and her main act was spirit control. Like Florence Cook, she too claimed to be able to harness the spirit of Katie King, along with other lesser-known entities. This made Holmes an ideal case study for Tylor, who was skeptical of the reports in London claiming that mediums could produce genuine, fully formed spirit manifestations.[77] Tylor pasted an advertisement for Holmes's spiritualist performance in his notebook (see fig. 2.1).

The advertisement encouraged "strangers, investigators, and non-believers" alike to attend Holmes's performances, and scrutinize the phenomena witnessed. This invitation was key to establishing Holmes's reputation as a genuine medium. It was a proclamation that she had nothing to hide. Moreover, the statement that she had performed similar acts all over the United States, Canada, and Central America, with glowing endorsements from "the most prominent gentlemen," added further credibility to her standing as an authentic medium.

Tylor had a rather low opinion of Holmes, and he painted a fairly unflattering picture of her in his notebook on spiritualism. His discriminatory comments exemplify his obviously biased views on gender and class. The juxtaposition between Tylor, the upper-middle-class British gentleman of science, and Holmes, the apparently lower-class American female medium, captures the often unbalanced power dynamics of many spirit investigations.[78] He described her as "a stout pasty-faced half-educated American with a black bush of curls." This sort of derogatory description was no offhand remark; it was part of his strategy for delegitimizing Holmes's status as a genuine psychic. As Stocking argued, "Tylor was inclined to make rather deprecating remarks about the appearance, the character, and the style of mediums . . . as a way of dismissing phenomena he could not adequately explain." Even before the performance had begun Tylor was preparing for his rejection of Holmes's mediumistic abilities. In this case her supposed lack of education in particular was crucial in undermining her trustworthiness.[79] Tylor also stated that Holmes's husband Nelson acted as the "master of ceremonies & looked after the fees."[80] This was a significant detail because it meant that Holmes always had an accomplice with her in the room.

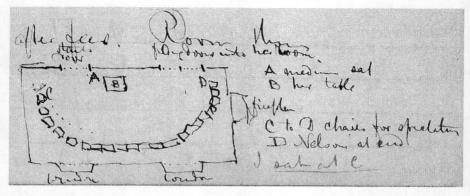

Figure 2.2. A diagram of the layout of Jennie Holmes's séance on November 4, 1872, at William Crookes's house from E. B. Tylor's notebook on spiritualism. *Source*: Edward Burnett Tylor, Notebook on Spiritualism, item 12, 7, box 3, Pitt Rivers Museum Manuscript Collection, University of Oxford.

The séance was rather poorly devised, according to Tylor, and from the onset he was skeptical of Holmes's professed powers. He discussed the layout of the séance in his notebook, stating that "the spectators were seated in a semi-circle; the medium sat at a separate table in the center."[81] Tylor included a small diagram of the séance's layout in his notebook (see fig. 2.2). He also described other preparatory arrangements that were made before the séance began: "On the table were rings, bells, etc., and tambourines & guitars about lying or standing. Mrs. H. first delivered an address, [and] then we were parted with the alleged view of putting each newcomer or skeptic between two oldcomers or believers. I was at the end."[82] Separating the believers from the skeptics was done to create a more balanced atmosphere, and inhibited the two camps from conspiring against one another during the séance. Several controls were also established in order to limit the chance of trickery occurring during the performance. Tylor noted that "Pieces of gummed paper were put to seal the two doors . . . [and] Lights were put out, I think the Medium having been first bound by someone present."[83]

A medium's believability was directly linked to how they framed their act. As Peter Lamont has argued, "it was important that the medium was seen as being neither responsible nor in control of what was happening."[84] Their credibility rested in the ability to eliminate any possibility of human intervention being responsible for the extraordinary phenomena observed. This was why Holmes had the doors to the séance room sealed and her hands bound. However, Tylor was not fooled by these so-called precautions. He suspected that Nelson, who had remained in the room unbound, was responsible for producing many of the manifestations occurring during the séance. Under the cover of darkness, he could easily manipulate the space.[85]

The main part of Holmes's spiritualist performance was her spirit control act. On this occasion Holmes claimed to be channeling a succession of different spirits, but the first to appear was a young Indigenous North American child named Rosie. According to Tylor, the spirit claimed to be able to see what the audience members were doing in the dark, and chastised him "for keeping one leg crossed over the other."[86] Tylor believed that it was conceivable for Holmes to have noticed his habit of crossing his legs before the lights were turned off. He also noted that if indeed the spirit was able to see the actions of the audience members in the dark, then she should have noticed that his free hand was "occupied in making the long nose for a good time in her direction."[87] The truthfulness of the spirit's claim was therefore brought into question. Yet this inconsistency in itself was not enough to discredit Holmes's performance as a whole. Tylor required more evidence if he were to dismiss fully the medium as duplicitous.

Tylor identified other flaws and discrepancies in Holmes's performance. He wrote that when he was touched on his head by an alleged spirit hand, "The contact . . . seemed to prove that the touch was by a rod or ruler." The cause of the so-called spirit manifestation appeared to have a rational explanation; one that could feasibly be human-made. He resolved that Nelson was culpable for producing the supposed spirit hand. Tylor had sat directly across from Nelson during the performance, putting him within easy reach of a stick or rod.[88] Tylor also noticed other flaws in Holmes's performance. In his notebook he wrote that while Holmes was being possessed by a spirit named "Irish Ann," her "simulated voice failed," and "she began *chi* in [an] American vowel but caught herself and harked back to put the word into Irish."[89] This fluctuation between accents further raised Tylor's doubts, and confirmed in his mind that the whole act was deceitful.

After the séance, Tylor walked back to his lodgings with Sir Wyke Bayliss (1835–1906), a well-known British poet and painter of cathedrals. He had also attended Holmes's séance at Crookes's house that evening, and agreed with much of Tylor's assessment. He too was unconvinced by the controls that were put in place by the medium, and suspected that Nelson was aiding his wife during the séance by producing the various spirit phenomena that they had witnessed. As was the case with Tylor, Bayliss also noticed inconsistencies in the alleged spirit voices that Holmes was channeling during the séance. After discussing these matters at length with Bayliss, Tylor concluded that the whole act was spurious, and he resolved that it was "the most shameful and shameless" spiritualist performance he had ever observed.[90]

The following night Tylor attended another séance, but this time at the house of Serjeant Cox.[91] The Crookes brothers were both in attendance, along with Cox's sister Mrs. Jaquet. A spiritualist healer known as Mrs. Olive led the séance.[92] As Alex Owen has discussed, many Victorian spiritualist healers advertised their services in the periodical press. Mrs. Olive was one of the better-known figures, and her specialty was mesmeric (or magnetic) therapy. Her business was so successful that she held two weekly sessions for free. Not only did this gesture demonstrate that Mrs. Olive was willing to share her so-called gift with people of all classes—including those who could not normally afford her services—but it also gave the impression that she was not simply using her mediumistic powers for profit.[93] This was all part of her strategy for constructing a reputable image of herself as a genuine medium. Prior to her conversion to spiritualism, Mrs. Olive lived in squalor. She came to realize her powers only after the death of her child, who, she claimed, visited her only a few hours after passing away.[94] In addition to her mesmeric healing abilities, Mrs. Olive also claimed to be proficient in spirit communication and spirit control. Tylor pasted a copy of an advertisement for Mrs. Olive's spiritualist performances into his notebook (see fig. 2.3).

From the beginning, Tylor seemed to be skeptical of Mrs. Olive's alleged mediumistic powers, and his doubts grew as the performance went on. Like Holmes, she began her séance by channeling the spirit of a young Indigenous North American girl, this one named Sunshine. Tylor noted that summoning the spirits of Indigenous North American children was a common occurrence in spiritualist performances. He attributed this trend to the fact that the movement originated in the United States, and suggested

▶ Figure 2.3. An advertisement for Mrs. Olive's spiritual services from an unreferenced periodical pasted into E. B. Tylor's notebook on spiritualism from 1872. *Source*: Edward Burnett Tylor, Notebook on Spiritualism, item 12, 7, box 3, Pitt Rivers Museum Manuscript Collection, University of Oxford. Although Tylor did not reference the source of this advertisement in his notebook it likely came from the following spiritualist periodical entry: "Mrs. Olive, Trance Medium," *Medium and Daybreak* 3 (1872): 436.

17

Nov. 5. 72 At Sergeant Cox's
36 Russell Square, present his sister
Mrs Jaquet, Mr & Mrs Walter Crookes,
to see the medium Mr Olive

After the performance her account of
herself was that her husband had
been a chemist at Devonport?)
and that Frank Herne (the courses
of the two mediums, Herne & Williams)
was much at her house when falling
into his medium-state & out of work
& hungry. Her own baby died & the
two it alive an hour after, which
was the means of her falling into
the spiritual profession becoming
so what a medium.
On sitting round the table with

that British-based mediums were likely emulating this popular practice because of its success among American audiences.[95] Mrs. Olive then claimed
to be possessed by the spirit of Franz Mesmer (1734–1815). No specific details
about Mesmer's life were mentioned during the possession, raising suspicions in Tylor's mind about the legitimacy of Mrs. Olive's mediumistic
powers. Tylor wrote in his notebook that "there was nothing to show that
she had any knowledge of what Mesmer was even like or thought or said."[96]

After the performance, when Mrs. Olive had departed, Tylor spoke with
Cox and Jaquet about the spirit manifestations that they had witnessed
during the séance. The three of them agreed that it was plausible that
Mrs. Olive had genuinely entered into a trancelike state. Moreover, Tylor
suspected that Mrs. Olive believed in "her own foolish imaginations." He
concluded that he did not think that Mrs. Olive was trying to deceive sitters
for economic gain, but that her performance was a kind of "self-delusion."[97]
By contrast, both of the Crookes brothers were utterly convinced by the medium's performance. According to Tylor, Walter Crookes was so persuaded
by the phenomena that he had witnessed during previous interactions with
Mrs. Olive that he even ordered his wife to follow a diet prescribed by the
medium under a trance-induced state.[98]

Nevertheless, with so many holes in her knowledge when possessed
by prominent spirits such as Mesmer, Tylor believed that Mrs. Olive was
probably suffering from some sort of "hysteria," which he recorded in his
notebook.[99] Tylor used this diagnosis, a condition that men typically associated with extreme emotional responses in women, to further discredit
Mrs. Olive. In his view, after the loss of her child, she no longer possessed
a rational mind, and therefore was more likely to believe in the genuineness
of her supposed extraordinary powers.[100] This exposes another example of
Tylor's gendered prejudices toward female mediums, which underpinned his
observations and which he used to justify his skeptical conclusions. It reveals
the uneven power dynamics of his spirit investigation, which juxtaposed his
elite knowledge and rationalism as a gentleman of science with the supposed
irrational beliefs of a delusional female medium.[101]

The séance at Cox's house did not provide sufficient data to suggest that
the spirit hypothesis was true, and animism by extension false. Tylor's ethnographic investigation, therefore, continued. On the evening of November
8, 1872, Tylor attended a larger séance at the Burns Spiritual Institution on
Southampton Row, in Holborn, London, a short-lived private organization

founded in 1863 by the journalist and publisher James Burns (1835–1894). The Burns Spiritual Institution provided a regular program of spiritualist performances, and was a central hub for spiritualist activities in London. Before converting to spiritualism in the 1860s, Burns was an advocate for the temperance movement. With his newfound belief in the spirit hypothesis, however, he began publishing the halfpenny weekly the *Medium* in 1869. It was later merged with another spiritualist periodical titled *Daybreak*, and the newly amalgamated periodical continued its run under the name of the *Medium and Daybreak* until 1895. For much of the second half of the nineteenth century, this periodical was a main source of news for the modern spiritualist movement in Britain.[102] Interestingly, Tylor appeared to use this journal for identifying the spiritualist events that he attended during his visit to London in 1872. The advertisements that he pasted into his notebook are identical to entries found in an issue of the *Medium and Daybreak* from November 1, 1872, which was published three days prior to his visit to see Jennie Holmes at William Crookes's house.[103]

Tylor visited the Burns Spiritual Institute to see the medium Lottie Fowler (née Connolly) perform. It was the largest spiritualist event that he had attended, and he recorded in his notebook that there was a "curious assortment of 20 people" in the audience. Originally from the United States, Fowler specialized in spirit materializations and clairvoyance. She became so renowned for her mediumistic powers that court officials allegedly consulted her regularly during a dangerous period of ill health for Prince Albert Edward (1841–1910)—later King Edward VII.[104] Tylor was highly skeptical of Fowler, and early on during the séance he was already identifying flaws in her performance. When under the possession of a young German spirit named "Annie," Tylor wrote that "she went round the company stopping before each with eyes shut, and telling each of the spirits she saw behind him or her. Her descriptions were guessed wrong 4 times in 5, or 9 in 10, but cleverly shifted and made right by getting something out of the sitter." Essentially, Fowler used various types of misdirection and cold reading techniques to gather information from audience members on their deceased friends and relatives. This gave the impression that she was genuinely communicating with the dead. Yet as Tylor recorded in his journal, most of it was guesswork and only provided vague or nonspecific information. For example, Fowler might correctly guess that a sitter lost a sibling, "but the name[,] age and description [were] more or less wrong."[105] Tylor left the séance concluding that most of the so-called spirit phenomena he witnessed was either "tricked out or guessed in the course of talk," and he believed that Fowler was a complete fraud.[106] Once again the theory of animism maintained its veracity.

The following day, Tylor met with William (1792–1879) and Mary Howitt (1799–1888) at their home in Notting Hill, London. The Howitts were Quakers who converted to spiritualism in the late 1840s. Over the next two decades, they built a fairly close-knit circle of friends with whom they held private séances. William Howitt also published several important works on modern spiritualism, including his most famous book, *The History of the Supernatural in All Ages and Nations* (1863).[107] The Howitts had strong ties to most of the leading figures within the British spiritualist movement, and they directed Tylor toward Emily Kislington. According to Tylor, Kislington's upbringing was rather unpleasant, and it was not until she began attending séances regularly that her wellbeing improved. Kislington would eventually go on to become the secretary of the British National Association of Spiritualists in 1877, and cofound the British Theosophical Society in 1878. However, in 1872 she was already highly respected within the community.[108] Kislington invited Tylor to her home on November 18, 1872, to attend a séance led by the well-known voice medium Caroline Elizabeth Bassett.[109]

Tylor had already received two reports that Bassett was a fraud. The first was from an unnamed sitter whom he had met at Fowler's séance a few days earlier, and the other was from Alice Margaret Lane-Fox (1828–1910), wife of Tylor's good friend and colleague, Augustus Pitt Rivers (1827–1900). Because Tylor viewed Lane-Fox as a particularly reliable source, he took her accusation of Bassett's alleged trickery seriously. Nevertheless, he still wanted to investigate her mediumship firsthand before drawing any formal conclusions.[110] Also present at the séance was William Henry Harrison (1841–1897), editor of the newspaper the *Spiritualist*, which ran between 1869 and 1882. As was the case with the *Medium and Daybreak*, it was a main organ of communication for the spiritualist movement in Britain, making Harrison one of the better-connected and informed members of the community.

The séance began with some spirit rapping, and Tylor determined that the noises were coming from where the medium was sitting. Then the lights were put out and Bassett allegedly began summoning various spirits, whose voices were heard all about the room. Tylor quickly rationalized how it was possible for Bassett to create the spirit voice effect. He wrote that "the medium for the purpose of sending her own voice from another spot, merely leant back &

put her head back, I held the tips of my fingers [behind her head] while she was speaking naturally, & as she actually came back upon them, [I] removed them."[111] Contorting her body therefore created her voice manipulation. Tylor also easily explained other phenomena produced during the séance. For example, at one point a supposed spirit touched him on his shoulder, but as he was sitting next to Bassett, he assumed she was responsible for producing it. He therefore put his fingertips between himself and the medium to see if she was touching him in the dark, and he stated that he met "an arm in the same feel of dress . . . as the medium wore."[112] He concluded that Bassett was a complete swindler, and he recorded in his notebook that "she cheated unscrupulously" during her performance.[113] He did not believe that Bassett was self-delusional, as was the case with some of the other mediums that he investigated, but an impostor trying to exploit people for economic and social gain.

Other peculiar incidents occurred later that evening. At dinner Tylor noticed that Kislington began acting strangely. She soon entered a supposed trancelike state, and Tylor wrote that "she was possessed by spirits but as she said they were influencing her to mesmerise her for her good." He continued by noting that "she talked entirely as [if] she believed herself possessed."[114] He reasoned that Kislington was likely suffering from some form of "hysteria," and this explained why she believed in the reality of spirit phenomena. As Judith Walkowitz has argued, framing female spiritualists as "maniacs" suffering from some form of psychosis was fairly common practice among Victorian scientists.[115] Tylor's remarks were therefore rather typical for a skeptic, and representative of the sorts of politics inherent in his writing throughout his notebook. It was a way to discredit believers of the spirit hypothesis, and it formed a key part of his strategy for strengthening the veracity of his visual epistemology.[116]

While walking back to his lodgings later that evening, Tylor spoke at length with Harrison. He told Harrison that he suspected Bassett of cheating because he had felt her movements during the séance, but Harrison countered by arguing that the "hands & arms & dresses that the spirits materialize are so like the ordinary ones that it's impossible to distinguish them." The movements that Tylor felt, according to Harrison, were most likely caused by spirit manifestations. The textural similarities between the spirit's dress and that of Bassett was caused by the spirit trying to emulate the medium's form—a common feature in so-called spirit materializations. Harrison added that "you scientific men must of course give natural explanations as seems sufficient to you."[117] Yet, according to Harrison, spiritualism could not simply be rationalized in this manner. Spirits lived beyond the natural world, and therefore did not conform to the same laws as other phenomena. The conversation ended soon after, and the two men parted ways. Despite this disagreement with Harrison, none of the phenomena he witnessed during Bassett's performance seemed authentic to Tylor, and he maintained his commitment to the theory of animism.

The following evening Tylor visited the home of a physician named George Bird, who was an acquaintance of Kislington. Bird was hosting a séance led by the world-famous American medium, Kate Fox.[118] Another sitter at the séance was the physician and pharmacologist Sydney Ringer (1835–1910), who worked at the University College Hospital in London. Ringer had previously participated in a sitting led by Fox, and he seemed to be convinced that her psychic powers were real. Tylor described Fox as "a nervous little woman." He recorded in his notebook that "everybody takes trouble to please her & keep her in good temper, for a little sets her flighty excitable hysterical little mind wrong."[119] Tylor was already framing Fox as possessing some sort of psychosis, and he was highly skeptical of her professed mediumistic powers. However, his skepticism soon waned. Before the séance began, Tylor spoke with Fox by the fire, and he said that spirit raps could be heard all about the room. He wrote, "They are strange bandings and creaking, some of them seeming to require much actual power."[120] Tylor was rather surprised by the amount of force produced by the various spirit phenomena he was hearing in the parlor room. He stated that "we went near the door, K.F. touch[ed] it with her finger tips producing loud thumps."[121] Tylor was intrigued by these sounds and struggled to explain how they were produced.

The group sat around a table and the séance began. More raps and knocks were heard almost immediately. According to Tylor, one spirit in particular, which identified itself as Elizabeth, interacted with the sitters for quite a while, rapping out messages and answering questions. However, much of these communications were semi-comprehensible. Tylor then placed a small slate frame with blotting paper attached to it on the table for the spirits to write on. The sitters held hands and soon the frame was flung off the table. Afterward, Tylor examined the object and he recorded in his notebook that "the 4 drawing pins which had attached the blotting paper to its corners were thrown near Dr. Ringer. When examined in the light it proved that the pins had been carefully extracted & the blotting paper was

Figure 2.4. A photograph of the Fox sisters fastened into E. B. Tylor's notebook on spiritualism from 1872. Margaret is on the left, Kate is in the center, and Leah is on the right. *Source*: Edward Burnett Tylor, Notebook on Spiritualism, n.p., item 12, 7, box 3, Pitt Rivers Museum Manuscript Collection, University of Oxford. The photograph was taken at M. E. Johnstone studios. "The Fox Sisters," catalogue no. 2009.148.2, Pitt Rivers Museum Manuscript Collection, University of Oxford.

left untouched somewhere else."[122] This discovery seemed to confound Tylor, and he was unable to explain decisively how the feat had been produced. He wrote, "There was much uncertainty in my mind as to whether we had properly joined hands at the moment when the frame was flung off the table."[123]

Reflecting on the events of Fox's séance later, Tylor had further difficulties making sense of what he had witnessed. He wrote in his notebook, "Altogether my experience of Kate Fox is very curious & her feats [are] puzzling to me." He continued, "Last night for the first time I saw & heard what deserves further looking into if I can get the chance."[124] Tylor may have begun the evening as a skeptic of Fox, believing that she was suffering from some form of hysteria, but after observing several unexplainable phenomena, he left the séance feeling baffled. For the first time in the course of his investigations on modern spiritualism, Tylor began questioning the foundation of his anthropological theories. Animism alone seemed insufficient in dismissing the phenomena produced by Fox, and in the absence of any other naturalistic explanation, the spirit hypothesis seemed to be gaining some impetus in his mind. This remark is also an implicit acknowledgment of how a medium was challenging Tylor's ideas, and forcing him to rethink the crux of his argument. Fox's inexplicable feats, such the apparent spiritual manipulation of the blotting paper, provide suggestive examples of the agency spiritualists held in shaping psychic investigations. Fox was not merely a passive subject in Tylor's study but an active contributor to his research. Interestingly, Tylor even fastened into his notebook a photograph of Kate Fox and her sisters Margaret and Leah (see fig. 2.4).

The last medium that Tylor investigated during his trip to London was the Reverend William Stainton Moses (1839–1892), whom he met through Cox. Moses originally trained as an Anglican clergyman at Oxford, graduating with an MA in 1863. Due to poor health, he was forced to change careers, and by 1871 he had been appointed as an English tutor at University College, London. He was introduced to spiritualism through a chance encounter with Fowler. Before realizing his own psychic powers, Moses had been investigating spirit phenomena as a skeptic. However, after experiencing some unexplainable occurrences, he gravitated toward the spiritualist cause, becoming an ardent proponent of the movement, and one of the most trusted and respected mediums in Britain. That said, he led only small private séances, and he usually refused to be tested by psychic investigators.[125] On first meeting him, Tylor recorded that Moses seemed "a plain good sort," and that he was "not only honest but not morbid-minded."[126] This was a very different characterization from how he framed female mediums such as Holmes and Mrs. Olive in his spiritualism notebook. Moses was not a delusional, hysterical woman, as Tylor had decided was the case with the others he met, but a seemingly rational and trustworthy man. Tylor even went so far to argue that Moses was a reliable confidant who could impart him with great knowledge of the modern spiritualist movement. Thus, Moses was positioned as an important voice of spiritualism and key informant for Tylor's psychical research.

On November 23, 1872, Tylor again visited the home of Cox to participate in a séance led by Moses. They sat in the library in a darkened room around a table. Soon various kinds of raps and knocks were heard, and a particularly pronounced tap was identified as being produced by a spirit known as Imperator.[127] This spirit featured heavily in Moses's writings on spiritualism, and was discussed at length in his later book *Spirit Teachings*, from 1883. According to Moses, "Imperator" was his "controlling spirit," and the main means by which he communicated with the spirit world.[128] Despite some supposed spiritual activity, Tylor viewed the affair overall as a failure. Moses was suffering from tuberculosis, and he was struck by coughing fits throughout the sitting. According to Tylor, this may have accounted for why there was a lack of spirit phenomena during the séance.[129]

The following morning, Tylor and Moses had a walk around the grounds of Cox's home. They discussed at length spirit photographs, and Moses explained some of the ways in which images were forged. It was another example of how Tylor was gaining valuable information from his interactions with spiritualists, who were becoming increasingly significant sources of knowledge for his investigation. The conversation then changed to the events from the evening before, and Tylor expressed his disappointment over the lack of spirit activity. Moses informed Tylor that when he left the room briefly there was a sudden influx in the spirit phenomena witnessed, and he attributed this shift in activity to a latent, but undeveloped, psychic power in Tylor that was "absorbing all the force" in the room.[130] Later that day, some interesting developments occurred when the group held another séance. Tylor found that he was passively falling prey to the performance of the séance. He wrote, "to myself I seemed partly under a drowsy influence, and partly consciously shamming, a curious state of mind which I have felt before & which is very likely the incipient state of hysterical simulation."[131] Essentially, Tylor believed that one reason why he (and other séance sitters) bought into the proceedings of spiritualist performances was that they were under a sort of group hypnosis. This may have been a rational explanation for why people believed in the reality of spiritual forces.[132]

Research into topics such as hypnosis and disassociation emerged alongside the rise of psychical research in the nineteenth century. These perceived rational and medicalized explanations for "altered states of consciousness" were often used to delegitimize spiritualist claims. For example, already in the late 1840s, John Elliotson (1791–1868), a physician at University College Hospital, was experimenting with mesmerism as a way of attempting to produce phenomena similar to clairvoyance.[133] William Benjamin Carpenter extended these ideas about disassociation and hypnosis even further during the 1850s and again in the 1870s, relating them to supposed spiritualist trances.[134] Thus, Tylor's attempt to interpret his own trancelike reactions during the séance as resulting from a kind of communal hypnosis fit with pre-existing explanations for the likely causes of supposed spirit and psychic phenomena, which were typically used by skeptical researchers when discrediting spiritualism and mediumship in their investigations.[135]

Reflecting later on his experiences with Moses, Tylor recorded in notebook that much of the reverend's credibility as a medium came from his earnest character. He wrote, "the evidence so far depends on Moses's honesty, he being a gentleman & apparently sincere." Tylor continued by stating that during the séance "his trance seemed real, & he made out that he knew nothing of what he had done." This strengthened the claim that his psychic abilities were genuine. Most importantly, though, Moses did not accept payment for his mediumistic services, and he often took a financial loss traveling to people's homes in the countryside. This was further compounded by Moses's poor health. Why would he subject himself to such difficult work if there were no clear gains? All of these factors contributed to the outstanding reputation Moses had built for himself as a reputable spiritualist. His open criticism of other spiritualist figures, whom he identified as tricksters, raised his status even more. Ultimately, Tylor was unable to conclude whether or not Moses was a fraud, although he was convinced that Moses genuinely believed himself to have psychic powers. Once again, Tylor could not rationalize his experiences with Moses as positive evidence in favor of spiritualism, but it certainly left him with some doubts. Tylor could maintain the veracity of animism, although his confidence in the theory was slightly unsettled.

Tylor had a few more exchanges with Moses before leaving London and returning home on November 28, 1872. During the course of his investigations, he met with several alleged mediums, and collected ample ethnographic data to found his conclusions on. By participating directly in the activities of spiritualists in London, he was no longer an armchair theorist basing his ideas on the testimonies of others, but an experienced eyewitness of so-called spirit and psychic phenomena. This did much to strengthen his claim as an authority on the subject of spiritual belief. Moreover, through his interactions with other investigators such as Crookes and Cox, as well as mediums such as Moses and Kislington, Tylor's knowledge of the modern spiritualist movement transformed considerably. Some of the figures he met, such as Holmes and Mrs. Olive, were easily discredited through naturalistic or rational explanations. By contrast, other figures were trickier, and generated all sorts of questions that Tylor was unable to answer with certainty. His experiences with Fox and Home are obvious examples. Adding further intrigue was that by the end of his investigations Tylor was becoming increasingly receptive to the proceedings of séances, even performing as though he was under some sort of trancelike state on one occasion. Tylor had much to consider on his journey home.

Armchair Reflections and New Directions in Animistic Thinking

Back at his home in Wellington, Somerset, Tylor reflected further on his ethnographic experiences in London, ruminating over whether he had seen any genuine spirit manifestations. He wrote in his notebook, "What I have seen & heard fails to convince me that there is a genuine residue. It all might have been legerdemain, & was so in great measure." But that seemed like an insufficient explanation for Tylor, who remarked that it was "too complimentary for the clumsiness of many of the obvious imposters" that he had met. The argument in favor of the spirit hypothesis seemed weak, in his view, and relied heavily on the "testimony of other witnesses."[136] For Tylor, the only evidence that would convince him of the reality of spirit phenomena was that which he collected himself. There may very well be genuine instances of supernaturalism, but until he witnessed it firsthand, it was all hearsay. He wrote, "I admit a prima facie case on evidence, & will not deny that there may be psychic force causing raps, movements, levitations[,] etc., but it has not proved itself by evidence of my sense, and I distinctly think the case weaker than written documents led me to think."[137] His verdict was less confident than before he embarked on his fieldwork trip, but he still remained committed to his skepticism of spirit and psychic forces.

Although Tylor's foray into psychical research has often been dismissed as little more than an amusing anecdote to his more serious anthropological research, his interest in spiritualism was intricately tied to his research on the anthropology of religion, and for much of the 1860s and 1870s, he had been developing his most important and long-lasting contribution to the discipline—his theory of animism.[138] Spiritualism posed a formidable threat to his anthropological research program because it it challenged the very foundation of Tylor's cultural paradigm. If animism was wrong the impact on the emerging discipline of anthropology would be significant, and there would be no guiding principle for interpreting human belief. Spirits would no longer be a cultural conception that was a survival of primitive thought, but evidence for an existence after death. That would mean that there was far more substance to the religious belief systems that Tylor was attempting to naturalize in his writings. Challenging the legitimacy of the spirit hypothesis was therefore an important test in demonstrating the validity of his work, and the explanatory powers of scientific naturalism.

Tylor's construction of animism transformed as a result of his investigations into spiritualism. As he sought to discredit spiritualists, he developed other means of rationalizing and explaining the causes that produced these extraordinary phenomena. The most common rationalizations were that the mediums were either cheats, delusional, or suffering from some form of "hysteria."[139] That did not mean that traditional animism, with its emphasis on superstition, was wrong, only that there were other ways to discredit the claims of spiritualists.[140] Someone could very well believe in the spirit hypothesis because of an inherent cultural survival of primitive thought, but if they demonstrated what he believed to be symptoms of mental illness, that was sufficient in itself for Tylor to disregard their testimony. By contrast, in cases where a medium seemed more honest, and of "sound mind," Tylor could always fall back on his theory of animism, which seemed to be the case with his interactions with Moses. Thus, Tylor was strategic in his application of classic animistic theory. His time in London did not lead to an overhaul of his anthropological theories, and it is significant that he did not publish the findings from his ethnographic investigations of psychics. The reasons are unclear, but perhaps Tylor did not feel that his fieldwork unearthed anything of note, and as such, it was unnecessary for him to defend his ideas further. Nothing that he had experienced undermined the veracity of his argument. Animism (broadly construed) continued forward as a viable theory within the discipline. Yet, as we will see in the next chapter, other interpretations of animistic belief developed, especially when it came to understanding modern spiritualism. Tylor might have held faith in his theories, but other anthropological figures during the late Victorian era were less certain.

3

Andrew Lang

The Revisionist

In 1913 the French historian and archaeologist Salomon Reinach (1858–1932) wrote a glowing obituary for the recently deceased Victorian polymath Andrew Lang. Over the course of his long career, Lang had published extensively on an array of topics ranging from literary criticism and Greek philosophy to Scottish history and second sight.[1] However, one of his most enduring legacies was his contribution to the anthropology of religion. Reinach wrote that "Lang was to Prof. Tylor what Huxley was to Darwin [in] that his chief and more lasting work consisted in popularizing the views of his master."[2] In many respects this characterization of Lang and Edward Burnett Tylor's relationship is accurate. During the 1880s Lang did much to promote Tylor's theories of animism and cultural survivals to a broad audience, by publishing articles in the general periodical press, and books such as *Myth, Ritual, and Religion* (1887).[3] However, all that changed in the 1890s as Lang's interest in psychical research grew.

Like many Victorians growing up in the decades following the publication of Charles Darwin's *Origin of Species* (1859), Lang experienced a kind of "cognitive dissonance" in which his spiritualistic and rationalistic beliefs clashed, leading to much intellectual headache.[4] The rise of secularism and materialism in Victorian Britain conflicted with his sentimental and romantic view of the world, and he increasingly mourned the loss of traditional folk culture and magic. Psychical research, with its commitment to studying the curious and unusual, had the potential to salvage any last drop of the marvelous that remained in modern society. Thus, Lang recognized its value for potentially sustaining his enchanted vision of the world.[5] However, as we have seen in previous chapters, many leading figures in Victorian Britain did not view investigations of spirit and psychic phenomena as a legitimate scientific pursuit, and for psychical research to gain some cultural authority, it had to appropriate theories and methods from the more established human sciences such as anthropology. The amalgamation of these two research fields was an uncomfortable one, and there was an obvious tension: psychical research, which employed unconventional scientific theories, such as telepathy, was incommensurable with the naturalistic principles of Tylorian anthropology.[6] If Lang wanted to marry the two fields, it was going to be a struggle.

In the preface to his book *Cock Lane and Common Sense* (1894), Lang remarked that "the testimony to abnormal events is much on a par with that for anthropological details, manners and customs," and continued, "The coincidence of report, in all ages, and countries, and from all manner of independent observers, unaware of each other's existence, was a strong proof of [their] general accuracy."[7] This concurrence of reporting suggested to Lang that perhaps not every instance of spirit and psychic phenomena could be accounted for by Tylor's theory of animism, which attempted to rationalize these occurrences through charges of fakery, superstition, and hallucination. The evidence did not stack up for Lang, and he began questioning the foundation of Tylorian anthropology. Was it possible that scholars were far too quick in dismissing cases of spiritualism, without giving them proper consideration? Was it also conceivable that skeptics could not fully refute claims of spiritualism and mediumship because they were dealing with abnormal facts that ran counter to conventional science? Was there something else at work that might be responsible for the continued presence of spirits and psychic forces? These were the sorts of questions that lingered in Lang's mind, and led to his critical re-evaluation of Tylor's ideas.

For animism to hold true, it had to be tested, and to do this, Lang went back to the sources. Whether spirits and psychics were to be confirmed as real or dismissed as relics of primitive superstitious thought, the credibility of the witnesses observing these extraordinary occurrences had to be established. Lang did not fully abandon his support for animism and cultural survivals, but he did provide a revisionist account of them. There was no doubt that some cases of spiritualism and mediumship were the result of trickery, super-stition, or delusion, but perhaps other instances could be explained through more idiosyncratic theories such as telepathy or thought transference. If we reflect on Lang's appropriation of animistic theory from his writings on spiri-tualism, Reinach's characterization of his relationship with Tylor seems rather mistaken. It is more fitting to argue that Lang was to Tylor what Wallace was to Darwin—a supporter of his master's work so long as it did not impede on the possibility of some hitherto unknown existence. If science were to expand its boundaries further, and prod deeper into the wondrous and mysterious, a hidden truth might be revealed—one that followed cogent principles like any other fact in nature. Lang, therefore, seems to have occupied a sort of middle ground between Wallace's position as a firm believer in the spirit hypothesis, and that of Tylor, who was a skeptic of spirits and psychic forces.

Cock Lane and Common Sense is Lang's most significant work on the intersection of anthropology and spiritualism. The book is an exercise in historical reconstruction that combines ideas and approaches from anthro-pology and psychology. It is historical because it includes data from the classical period to the nineteenth century. It is anthropological because it uses ethnographic data from all over the world. It is psychological because it deals with matters of the mind, such as hallucination theories, conceptions of thought transference, and telepathy. At its foundation is the comparative method. There was an overwhelming amount of evidence professing to have documented genuine spirit and psychic phenomena. If there was any weight to these claims, Lang believed, then they could be thoroughly scrutinized and still hold true. Similar cases had to be compared for consistency, and the more the reports had in common, the more likely there was an element of legitimacy in their claims. Each observer also needed to be evaluated on a case-by-case basis. Were they skilled observers with training in science or medicine? Did they possess an honest and forthright character? Were they philosophically minded? All of these aspects needed to be weighed before a report was deemed as trustworthy.

Although Lang is a familiar figure to most scholars interested in Victorian anthropology, folkloric studies, and psychical research, he has a marginal foothold in the secondary literature. This is somewhat surprising, given that he published a staggering number of books and articles during his long career, and contributed energetically to the disciplinary development of multiple fields.[8] Only one short biography has been published on him, and it was released in the 1940s.[9] His amateur status as a researcher working outside of the academy certainly had a role to play in limiting his standing within the discipline. His association with psychical research, combined with his challenge to animism, played a key part as well. For example, reflecting on some of his intellectual quarrels with Tylor in 1907, he stated, "He who would vary from Mr. Tylor's ideas must do so in fear and trembling (as the present writer knows from experience)."[10] *Cock Lane and Common Sense* posed a genuine problem for the more hardline animists, because if concepts such as telepathy were proven to be real, the foundation of Tylorian anthropology would crumble. So-called primitive people would no longer be seen as misinterpreting extraordinary phenomena in ancient times but as recording an important archive of psychic phenomena. There was much at stake, and Lang's thoughtful and detailed study of spiritualism requires full consideration. The evidence was there, but was the interpretation of it justified? The gauntlet had been thrown and Lang provided a sound case for potentially re-evaluating the theoretical limits of the discipline.[11]

Microhistory provides a useful analytical model for understanding how Lang interpreted his evidence and positioned himself as an authority in the study of spiritualism.[12] Through a detailed examination of the three key case studies from *Cock Lane and Common Sense*, it is possible to reconstruct Lang's observational practice. This in turn allows us to understand his revisionist model of animism. In each case, Lang carefully assessed the credibility of the evidence available to determine whether there was any substance to the claims of extraordinary phenomena. If he could show that credulity, superstition, and delusion were unable to fully explain the occurrences that were observed, telepathy, or some other unseen force, had to be taken more seriously. He believed that psychical research needed to be given a fair hearing, and the scientific community had to recognize its legitimacy. By getting scientists to accept psychic forces as genuine, Lang sought to salvage his romantic vision of a premodern, enchanted world.

Andrew Lang and the Making of a
Psycho-Folklorist and Anthropologist

Lang was born in 1844 in Selkirk in the Scottish Borders and came from a middling-sort family. His father was John Lang (1812–1869), a sheriff clerk of Selkirkshire, and his mother was Jane Plenderleath Sellar (1821–1869), daughter of Patrick Sellar (1780–1851), the factor to George Granville Leveson-Gower (1758–1833), the First Duke of Sutherland. From a young age, Lang was highly studious, and as George Stocking recounts, "at five [he] foreshadowed the reading habits of his later years, by setting six books upon six chairs and reading from one to another as his interest waned."[13] He attended both Selkirk Grammar School and the Edinburgh Academy, where he was a bit of an outsider, struggling to make friends, and spending much of his leisure time buried in books on old fables and legends. By 1861 he had gained admittance to the University of St Andrews, where he studied more seriously contemporary scholarship on folklore and mythology. It was also during this period that his interest in occultism blossomed, and during his spare time he allegedly dabbled in necromancy and alchemy. His time at St Andrews was short-lived, however, and he soon transferred to the University of Glasgow to seek new opportunities. After spending a year there, he qualified for a Snell Exhibition (a scholarship), which allowed him to undertake further studies in classics and literature at Balliol College, University of Oxford, where he was awarded a first-class degree in 1868. He became a fellow of Merton College that same year—a position he held until 1875 when he married Lenora Blanche Alleyne (1851–1933). For the remainder of their lives together, the couple split their time between living in London and in St Andrews.[14]

The quality of Lang's scholarship transformed considerably at Balliol under the tutelage of renowned figures such as the theologian and classicist Benjamin Jowett (1817–1893) and the philosopher and political reformer Thomas Hill Green (1836–1882). Jowett introduced Lang to ancient Greek texts by figures such as Plato (428/27–348/47 BCE) and Thucydides (460–400 BCE), while Green taught him about Hegelian metaphysical historicism. These topics would influence Lang's later writings on spiritualism, shaping both the theoretical foundation and evidentiary standards of his research.[15] However, his love for folklore and mythology persisted, and through his incessant reading he stumbled upon the anthropological works of the Scottish lawyer John Ferguson McLennan (1827–1881) and Tylor.[16] During much of the nineteenth century, studies of folklore and mythology fell within the disciplinary sphere of anthropology, and these tales were important ethnographic sources for interpreting human cultures. McLennan's and Tylor's respective works were influential to Lang, because they underscored the value of using comparative methods for analyzing cultural phenomena that were steeped in allegorical meaning. Comparative analysis was crucial to Lang's later investigations of spiritualism and mediumship, and he even refers to it as the "anthropological method" in *Cock Lane and Common Sense*.[17]

Although Lang wrote prolifically across an array of subjects, anthropology came to dominate much of his academic research in the decades between the 1860s and 1880s.[18] He became an ardent follower of Tylorian anthropology, which he positioned in opposition to the philological model of Friedrich Max Müller (1823–1900) that was popular during the period.[19] So persuaded was he by the arguments in *Primitive Culture*, Lang dedicated his book *Custom and Myth* (1884) to Tylor. Animism in particular, continued to form the core of Lang's anthropological paradigm, and in *Myth, Ritual, and Religion* we see the most sustained application of Tylorian theory in his writings.[20] Using his skill in comparative studies, Lang systematically traced the form and meaning of myths, rituals, and religions across various cultures and historical periods, linking them to so-called primitive forms of superstition. This research served him well in his later studies of spirit and psychic phenomena, because he became highly proficient in verifying the credibility of personal testimonies for anthropological investigations.

Much like Tylor's visual epistemology, Lang's verification technique for establishing the credibility of his evidence had four main principles.[21] First, if observers possessed comprehensive training in a field that was considered requisite to anthropological investigations, such as medicine or natural history, they were considered to be credible witnesses. Similarly, if observers had a sound knowledge of subjects such as law, physics, or philosophy, they too were deemed trustworthy, because of their analytical and discerning minds. Second, if the account employed a type of collective empiricism, where multiple reports contained analogous information on the same objects, topics, events, or peoples, the observations reinforced the validity of one another. Third, if a witness could reinforce his or her claims with the observations of others, who had professed to see similar phenomena, this added further credibility to an account. Fourth, if multiple witnesses were present at the same incidents, and produced corresponding reports, they were all identified as trustworthy observers.[22]

Things began to change as the decade wore on, and by the mid-1880s Lang's confidence in animism waned, and this might be linked to the intellectual climate of the day. As Stocking has argued, Lang was part of a generation of scholars who experienced a kind of "rationalistic doubt" during the heyday of scientific naturalism, when secularism and materialism were on the ascent. For many of this generation growing up in the post-Darwinian era, there was a move back toward either traditional religious faith or more idiosyncratic beliefs such as spiritualism. It is possible that Lang's abrupt change of thought regarding animism was linked to some kind of cognitive dissonance or intellectual crisis that he was experiencing.[23] He began questioning whether anthropological discourse provided adequate explanations for the causes of spirit and psychic phenomena. As we have seen, animism offered three main rationales: spirits and psychic phenomena were the result of either trickery, superstition, or delusion. For Lang, however, this seemed too narrow, and he argued that the topic required more thorough consideration. What if Victorian writers were getting it wrong? What if an overcommitment to currently accepted natural laws blinded researchers to more aberrant possibilities? The problem might be that claims of spirit and psychic phenomena were being treated as typical facts of nature. Maybe spirits and mediums needed to be handled differently, as abnormal facts, that did not abide by known natural laws.[24] Lang did not fully abandon animistic theory, but he did want to provide a revisionist account of it.

In 1885 Lang published "The Comparative Study of Ghost Stories" in the popular periodical the *Nineteenth Century*.[25] It was his first notable attempt at trying to merge anthropology with psychical research. Lang argued that despite its relevance, anthropology had not given thoughtful consideration to the reports of witnesses claiming to have experienced genuine spirit and psychic forces. Researchers were far too quick to dismiss these accounts. However, the number of testimonies that had been collected in all parts of the world, and in all historical periods, suggested that there was good reason to take these claims more seriously. According to Lang, Tylorian anthropology in particular had not "paid very much attention to what we may call the actual ghost stories of savages—that is, the more or less well-authenticated cases in which savages have seen the ordinary ghost of modern society."[26] What was needed was a classification system with a basic set of criteria that allowed researchers not only to identify types of ghost stories but also to compare them and verify their authenticity.[27]

Lang contended that ghost stories were prevalent in all societies from "races as low as the Australians" to "contemporary European civilization."[28] He identified four main types. The first were stories in which a spirit (often the recently deceased) took the form of an animal—usually one that had sacramental importance to the culture. While this sort of story was most common among Indigenous peoples, Lang stated that it was also "prevalent and fashionable" among European peasantry. It was rarely seen among the so-called educated classes of Europe and North America. These types of ghostly encounters were often ascribed prophetic meaning.[29] The second type of story contained "professional ghosts" that were summoned by mediums and obedient to their commands. Examples of this sort of ghost story were collected all over the world, and Lang wrote, "These spirits, which come 'when you do call them,' behave in much the same manner, and perform the same sorts of antics or miracles, in Australian *gunyehs,* in Maori *pahs,* and at the exhibition of Mr. Sludge, or of the esoteric Buddhists."[30] These were the sorts of spirits that were usually described at séances, and came to be representative of the modern spiritualist movement. The third type of story involved the "nonprofessional ghost," which did not beckon to the calls of mediums, appeared unexpectedly, and was usually impassive to the living. Ghosts in this group haunted buildings, urban settings, or the countryside, and often manifested themselves around the time of death. When they did communicate it was mainly "for the purpose of warning friends of their own approaching decease."[31] Finally, there was a class of ghost stories in which the spirit was fairly inactive and did not engage in any sort of meaningful communication with the living. These ghosts were emblematic of classic hauntings.[32]

While classifying such ghost stories was an important step in better understanding the persistence of this cultural phenomenon in all ages and regions of the world, the big question still remained: why do these stories exist? Here Lang began critically evaluating the data. He believed that there were several possible explanations for why all cultures recorded ghost stories. First, it was conceivable that there was an "internal groundwork of fact" underlying all types of ghost stories, regardless of whether they came from savages or "civilized man." Second, if there was not some element of truth to these stories, then another explanation was that all humans were susceptible to similar types of "recurring hallucinations" that were inherited from their "untutored ancestors." Third, it was possible that some mythopoeic thought or false belief was being passed down to every generation. Fourth, it was likely that these stories were grounded in a misinterpretation of natural phenomena based on primitive, superstitious thought.[33] Anthropologists tended to either ignore or dismiss claims that attempted to verify the existence of spirits or psychic forces, but generally they supported explanations that rationalized ghost stories as the product of delusion, sickness, or superstition. The strength of anthropology was in its contextual and comparative approach to the subject; its weakness was an inherent skepticism to so-called supernaturalism.

By contrast, psychical research had its own problems when dealing with the testimonies of observers professing to have witnessed genuine spirit and psychic phenomena. Usually, these studies were overly focused on modern cases, and did not contextualize them sufficiently using historical or ethnographic records. Lang wrote, "The friends of psychical research have collected modern stories of the actual appearance of apparitions without paying much attention, as far as I am aware, to their parallels among the most backward races, or to their medieval and classical variants."[34] What was needed was an investigation that combined the contextual and comparative methods of anthropology with the earnestness of psychical research for the topic. In doing so, the veracity of the evidence could be more thoroughly established, and Lang contended, "Though this sketch of a new comparative science does not perhaps prove or disprove any psychical or mythological theory, it demonstrates that there is a good deal of human nature in man. From the Eskimo, Fuegians, Fijians, and Kurnai, to Homer, Henry More, Theocritus, and Lady Betty Cobb, we mortals are 'all in a tale,' and share coincident beliefs or delusions. What the value of the coincidence of testimony may be, how far it attests facts, how far it merely indicates the survival of savage conceptions, Mr. Tylor and Mr. Edmund Gurney may be left to decide."[35] To establish credible witnesses in studies of spiritualism broadly construed, the evidence needed to be tested, and that was possible only if researchers adopted a more omniscient and erudite approach to the matter. Anthropology and psychical research working together had the potential to resolve the great mystery of spiritualism in all cultures and ages. The remainder of Lang's essay provided examples of how to conduct this integrated approach with cases drawn from historical and ethnographic sources.

Lang's broad education and training in subjects ranging from classics, history, and literature to folklore, mythology, and anthropology made him well suited for a detailed and comparative study of spiritualism, which combined aspects from all of these research fields, and allowed him to assess the credibility of his sources more scrupulously. This laid the groundwork for a new kind of investigation of spirits and psychics that combined what he believed to be the best (and most relevant) aspects of these research fields, with a particular emphasis on anthropology and psychical research. "The Comparative Study of Ghost Stories" was the first step, and it was a compelling argument, but not an exhaustive one. Lang provided his readers with an introduction to the topic, but he needed more to make a really strong case for taking seriously his vision of the scientific study of spiritualism and mediumship. Further data needed to be amassed and rigorously compared, in order to prove that the coincidence of testimony of those professing to have observed spirit and psychic phenomena contained some underlying truth. What that truth was, Lang was still unsure, but whatever it proved to be, it would become fundamental to his future research. Over the next ten years he continued his efforts, and those labors formed the foundation of *Cock Lane and Common Sense*.

Weighing the Evidence

In the opening pages to *Cock Lane and Common Sense*, Lang made a strong case for taking his evidence seriously. It all came down to the trustworthiness of his sources. Had credible witnesses produced these accounts? He acknowledged that personal testimonies could be an unreliable basis for substantiating the credibility of a scientific inquiry, because all witnesses, regardless of their background and education, were susceptible to their inherent biases. However, if the "anthropological method" was used correctly, it was possible, in Lang's view, to verify the accuracy of the evidence. The researcher just needed to be cautious when assessing the data. He wrote, "The writer has answered that . . . several of the statements in this volume are not the chatter of personal paragraph-makers, but are, in some cases, attested on oath, in others, have survived strict cross-examination."[36] Even if the evidence had already undergone some level of critical assessment to confirm its legitimacy, there were still some notable issues for researchers to be wary of when weighing the value of the material.

Since the seventeenth century, researchers had been grappling with all sorts of techniques for determining whether a person's observations were credible. These included collective testimony, replication, and virtual re-creation through text.[37] Lang was following in this tradition of verification, and he explained how some observers were likely to include false information in their statements, either through misremembering the circumstances or through willful deceit. This was not exclusive to spirit investigations, but occurred in all sorts of research fields. It was the duty of the researcher to determine whether the information provided in a testimony was dependable. If it was evident that the witness was mistaken or dishonest, their account had to be dismissed. Lang contended that "in every field of study there occur incorrect descriptions, and, in all, the bias of witnesses has to be allowed for: illusions of memory, of vanity, of the artistic instinct have to be discounted."[38] It was not just believers and supporters of spiritualism who were predisposed to distorting their testimonies; skeptics, too, were prone to misrepresenting and exaggerating their findings.

A fairly high-profile example of a skeptic doctoring his conclusions was Sir David Brewster. At the invitation of Lord Henry Peter Brougham, Brewster attended a séance at the home of Serjeant Edward William Cox in 1855. Daniel Dunglas Home was the medium under investigation, and despite several attempts to expose how his extraordinary feats were produced, the investigators were left confounded. However, that was not what made this case significant among spirit investigators. What intrigued them was how even a skilled scientific observer such as Brewster could produce unreliable and tampered evidence. Within a few weeks of witnessing Home's incredible displays, he produced two conflicting reports: one for the general periodical press dismissing the genuineness of Home's so-called psychic powers, and another via private correspondence to his daughter, Margaret Maria Gordon (1823–1907), revealing that he was baffled by the phenomena he had witnessed.[39] "In which letter did Sir David tell the truth?" Lang wondered. "Here we have a trained observer, an authority in exact science, who flatly contradicts himself about an event not a week old at the time of his writing. He has one story for the public, a story directly contradictory of it for his family circle."[40]

Brewster was worried about his scientific reputation as an expert in human perception, and this influenced how he portrayed the investigation publicly. Among his many significant contributions to the physical sciences, his *Letters on Natural Magic* (1832) aimed to explicate the processes by which illusions were created. His inability to explain how Home produced his amazing feats undermined his authority as a psychical researcher, expert in optics, and staunch skeptic.[41] If he wanted to maintain any semblance of cultural or scientific authority, he had no choice but to publicly dismiss the legitimacy of Home's mediumship in the press, even if he was unsure how his feats were done. Brewster's indiscretions exemplified Lang's key point about why it was important to contextualize and critically evaluate personal testimonies—because witnesses had individual motives, which affected their reliability as objective observers.[42]

Another issue when assessing the validity of personal testimonies in spirit investigations was that there were a lot of swindlers in the spiritualist community. Were observers discerning enough to tell the difference between so-called real mediums and fake ones? There were plenty of examples where tricksters posed as genuine mediums and fabricated all sorts of supposed spirit and psychic phenomena in order to exploit people for social and financial gain. Investigators worked diligently to try and expose most of the standard phenomena occurring at séances as spurious. However, Lang still had some lingering doubts. Were any of these phenomena real? Was it possible that tricksters were copying genuine phenomena? Even if it were feasible to show that some of these occurrences were legitimate, Lang argued that spirit investigations would never receive a fair hearing among skeptics.

Since the eighteenth century, rationalism had been marginalizing earnest inquiry into spiritualism and mediumship as disingenuous. Those attesting to have witnessed authentic spirit or psychic phenomena tended to hide their experiences, fearing that they would be derided as credulous fools by their family, friends, and colleagues. Lang wrote, "Common-sense bullied several generations, till they were positively afraid to attest their own unusual experience. Then it was triumphantly proclaimed that no unusual experiences were ever attested. Even now many people dare not say what they believe about occurrences witnessed by their own senses."[43] Fearmongering was responsible for stacking the evidence in favor of skepticism, and Lang wanted to reverse this trend. It was only through a balanced and candid investigation of extraordinary phenomena that he believed a fair judgment could be formed, one that was free from the contempt of closed-minded rationalists.

Fanaticism was not just a problem among disbelievers, and Lang reminded his readers that there were just as many zealots among proponents of spiritualism. Like their skeptical counterparts, these ardent supporters damaged the reputation of spirit investigations by being far too uncritical of the evidence favoring the legitimacy of the spirit hypothesis. Take, for instance, the famous American spirit investigator Robert Dale Owen. His two biggest works on spiritualism, *Footfalls on the Boundary of Another World* and *The Debatable Land between This World and the Next*, were considered by many readers to be overly sympathetic toward the spiritualist cause, and did not provide balanced perspectives.[44] Lang wrote, "On the other side we have writers, like the late Mr. Dale Owen, who avow and display a potent bias in favour of establishing a belief in a future life. They, too, just like the sceptics, blink awkward facts."[45] A more reflexive approach that underscored the complexities of psychical research was needed.

The more it could be shown that spirits and psychics were observed in all cultures and across all ages, the easier it would be to make the case for treating spirit investigations more seriously. The uniformity of descriptions recorded in personal testimonies suggested to Lang that there was some hidden universal truth to be exposed. He stated, "It is desirable to know why independent witnesses, practically everywhere and always, tell the same tales. To examine the origin of these tales is not more 'superstitious' than to examine the origin of the religious and heroic mythologies of the world. It is, of course, easy to give both mythology and 'the science of spectres' the go by."[46] Spiritualism needed to be treated in the same way as folklore and mythology. The roots of these stories were grounded in something factual, but the actual meaning needed to be disentangled from the lore that surrounded them. Much like the more basic conceptions of animism, charges of superstition alone were too simplistic for explaining every instance of spiritualism or mediumship. There were other possibilities, but researchers needed to be open-minded enough to consider them.

It was certainly conceivable that most (if not all) of the personal testimonies with witnesses professing to have observed genuine spirit or psychic phenomena were mistaken, and that every example could be explained through superstition, fraud, or delusion. This was the stance that most proponents of Tylorian animism took. However, if all human groups were susceptible to the same miscomprehensions, that alone made the phenomena worthy of further investigation. Lang stated, "It is a question whether human folly would, everywhere and always, suffer from the same delusions, undergo the same hallucinations, and elaborate the same frauds."[47] He argued that it is not just among uneducated peoples that accounts from witnesses professing to have observed real spirit or psychic phenomena appeared. There were abundant examples among modern, educated people as well.[48] This added further weight to the argument that superstition or trickery could not explain away all instances of spirit or psychic phenomena. If discerning and educated minds experienced these extraordinary occurrences as well, credulity alone seemed an unlikely cause.

Investigations into the reality of spirits and psychic forces deserved thoughtful, scientific consideration, and according to Lang only through prolonged inquiries could a balanced and reliable opinion on the matter be formed. Locating a couple of examples of fraud was not enough to dismiss every testimony ever written. Nor was it sufficient to attend only a small number of séances with mixed results in order to confirm that the spirit hypothesis was wrong. For Lang, "This affair demands the close scrutiny of years, and the most patient and persevering experiment."[49] *Cock Lane and Common Sense* was exactly that—the product of decades of research. Not only had Lang attended numerous spiritualist performances but he meticulously combed through countless historical and ethnographic sources, locating examples of spirit and psychic phenomena. He determined that there were likely four main causes responsible for producing these extraordinary phenomena: superstition, hallucination, fraud, and telepathy. It was this fourth cause that distinguished Lang's version of animism from the standard Tylorian one, and it generated two key, connected questions: First, who could be trusted as a reliable witness? Second, what kind of observer was more suspect?

One observer whom Lang believed to be trustworthy was the German Jesuit theologian Peter Thyraeus (1546–1601), who had written extensively on hauntings, demonology, and exorcisms in the sixteenth century. His most famous book, *Loca Infesta* (1598), was a key source for early spirit investigations, and Lang studied it fastidiously.[50] Thyraeus used a kind of prototypical version of the "anthropological test of evidence" in his book to establish the credibility of his analysis, and through a process of historicizing his data, he was able to show how similar types of testimonies were documented recurrently since ancient times.[51] This coincidence of reporting was highly suggestive, and Thyraeus believed that it verified the existence of spirits—though he framed them as "sprites," "daemons," and "souls." *Loca*

Infesta also provided one of the earliest examples of a classification system for making sense of these entities, and the phenomena they produced.[52] For instance, a spirit haunting a home could produce three kinds of effects: visual, auditory, or touch. Although Thyraeus was a confirmed believer in the existence of spirits, Lang asserted that his book was reliable. Thyraeus had undertaken sustained research into the topic, possessed significant firsthand experience observing phenomena, and was highly critical of his secondhand sources, as he rigorously compared them to determine their veracity. So long as one recognized Thyraeus's biases, which were intertwined with his deep Catholic beliefs, his works could still be used effectively in studies of spirits and psychics.

Lang was far more critical toward the Society for Psychical Research's (SPR) investigation of alleged hauntings.[53] Since its formation in 1882, Lang had been actively involved in the development of the SPR's research program, holding close ties to some of its leading members, including Henry (1838–1900) and Eleanor (1845–1936) Sidgwick, Frederic W. H. Myers (1843–1901), Frank Podmore (1856–1910), and Edmund Gurney (1847–1888), to name a few. However, that did not stop Lang from problematizing the group's research practices. He explained that in 1884 members of the SPR identified nineteen stories that they deemed to be "first-class" examples of hauntings "based on good first-hand evidence."[54] A key concern was determining what were the characteristics of a genuine ghost story. Any case that deviated too much from these qualities was probably spurious, or a misinterpretation of the phenomena witnessed. With a strict set of characteristics necessary to confirm a ghost story as genuine, many cases were too easily rejected, and Lang believed that part of the problem was that the SPR's analysis of the material was poorly executed and based on false assumptions. For instance, the SPR investigators had "never yet hired a haunted house in which the sights and sounds continued during the tenancy of the curious observers."[55] It was therefore assumed that the earlier testimonials professing to have seen a ghost were probably mistaken. However, Lang argued that there was another reason why it was possible that the investigators did not observe any extraordinary phenomena—ghostly activity is erratic. It can come and go, and long periods of time can pass without any occurrences, only to return months (or even years) later.

Lang recalled one story in which a family took up residency in a new home, and soon began experiencing unusual phenomena. He stated that a ghost "made loud noises, it opened doors, turning the handle as the lady of the house walked about, it pulled her hair when she was in bed, plucked her dress, produced lights, and finally appeared visibly, a hag dressed in grey, to several persons." Then much time passed without any occurrences until one day it was "as noisy as ever, and appeared to a person who had not seen it before."[56] Because the new observer had no knowledge of the previous experiences, this added legitimacy to the story. Lang explained that this was not an isolated case, and that hauntings did not have to be constant. He wrote, "When we hear of a haunted house, we imagine that the ghost is always on view, or that he has a benefit night, at certain fixed dates, when you know where to have him. These conceptions are erroneous, and a house *may* be haunted, though nothing desirable occurs in presence of the committee."[57]

For psychical research to be done properly and fairly, Lang also argued that it was essential for investigatory teams to include a least one member who was sensitive to spirit and psychic phenomena. He wrote, "It is thought very likely that, where several people see an apparition simultaneously, the spiritual or psychical or imaginative 'impact' is addressed to one, and by him, or her (usually her) handed on to the rest of the society."[58] Lang contended that spirit and psychic forces mainly affected those with mediumistic abilities, but through them even skeptics could sense the phenomena. If no medium were present at an investigation, the chances of observing extraordinary occurrences diminished greatly. Because the SPR investigatory team did not include someone with this sort of psychic disposition, their results were negative and inconclusive. Lang remarked, "Now if the committee do not provide themselves with a good 'sensitive' comrade, what can they expect, but what they get, that is, nothing?"[59]

In the SPR records, a typical haunting included both auditory and visual phenomena. In the case of the former, Lang explained that "the sounds are footsteps, rustling of dresses, knocks, raps, heavy bangs, noises as of dragging heavy weights, and of disarranging heavy furniture. These occur freely, where nobody can testify to having *seen* anything spectral."[60] Visual phenomena differed and often took the form of phantasms, which were described as "figures beheld for a moment with 'the tail of the eye' or in going along a passage, or in entering a room where nobody is found, or standing beside a bed, perhaps in a kind of self-luminous condition." Combinations of these two forms of phenomena were typical of most modern hauntings, and were seen as essential characteristics for determining whether a location had a ghost. However, Lang argued that a witness could easily misconstrue noises and visual manifestations as spirit phenomena. For instance, "Noises may be naturally caused in very many ways: by winds, by rats, by boughs of trees, by water pipes, by birds."[61] To confirm these occurrences as spirit in origin, all possible naturalistic explanations had to be dismissed first. Then, by employing a form of "collective empiricism," through amassing reports from different hauntings professing to have experienced similar sensations and spectacles, strong patterns might emerge that added much weight to spiritual and psychical claims.[62]

After outlining for his readers the main issues that he had with the SPR's investigation of hauntings, Lang then scrutinized closely the details of five of their case studies. In each instance, Lang believed that the SPR misunderstood, overlooked, or distorted important details. Whether this was done purposefully or through carelessness is unclear, but either way, the implications were great. By omitting information, or ignoring significant facts, it was easier to reject or rationalize the occurrences as being the result of natural or mechanical causes. By critically reappraising the materials, Lang sought to rectify these shortcomings and foster a more thoughtful and rigorous study of spirits and psychics.[63] It was not that the SPR got it wrong per se, but their findings were based on numerous mistakes. If future researchers were more careful, and followed his introspective approach, Lang believed, a more nuanced analysis could be achieved, one that revealed a very different understanding of what an "authentic" haunting entailed.

The first case occurred at a small villa in an undisclosed location in the English countryside. The main witness was "Mrs. C," who kept seeing "a tall dark-haired man" in various rooms around the house. The appearances then stopped for two months before the apparition was seen again in her drawing room. At this point Mrs. C informed her husband of what she had been witnessing, but nothing more came of it until other members of the family began seeing the figure too. It was described in a similar manner by the children, and was heard saying in a deep and sorrowful voice: "I can't find it." The concurrent testimonies of the family members added weight to the case. It seemed as though for several months, they were unaware of what each other had been witnessing, and yet the details of their accounts still matched.[64] What was most important to Lang was that this was an exemplary case of a supposed ghost shifting between heightened activity and idleness for months at a time. If an investigator were to spend only a short period of time at the home to inspect the circumstances, it was possible that they would not witness any strange phenomena. Thus, the example supported Lang's key point that investigators needed to reside at haunted locations for prolonged periods of time before dismissing testimonies as erroneous.

The second case gave a similar story, and Lang recounted that "some ladies and servants in a house in Hyde Park Place, see at intervals a phantom housemaid." The ghost's activity was unpredictable and periods of time could pass without any disturbances. This case again highlighted why a more sustained investigation into hauntings was required before dismissing them.[65] There were other issues with the SPR's analysis of case files as well, and Lang recalled a third example where the main issue was not the erratic activity of the so-called spirit but the time of day when it was seen. A father and his young son living in India saw the figure of a "Hindoo native woman" separately and on different occasions in their house. The child's observations occurred during a semi-conscious state after being awoken from his sleep.[66] This was an important detail, and possibly accounted for the sudden appearance of the spirit. It was conceivable that the child's ghostly encounter was the product of a dream-induced hallucination—an explanation fitting with Tylorian animism. It was during these liminal states that the mind was easily fooled. When someone was half awake and half dreaming, unexpected visions seemed all the more real.[67]

Lang included a detailed discussion of different types of hallucinations in *Cock Lane and Common Sense*, and in many respects, it brought the theoretical foundation of his work closer to that of Tylor in *Primitive Culture*. The main difference was that Lang sought to align his book more closely to the current theories and practices of psychical researchers, such as those of Podmore, Myers, and Gurney.[68] Lang argued that there were two broad categories of hallucinations: "Those [in] which the percipient (or percipients) believes, at the moment, and perhaps later, to be real; and those which his judgment pronounces to be *false*."[69] In both cases the causes were usually linked to either illness (cognitive and physical) or intoxication. Hallucinations originating from some sort of mental process were usually identified as "hallucinations from within," and Lang provided some examples, such as cases in which "a patient begins with a hallucination of the intellect, as that inquisitors are plotting to catch him, or witches to enchant him, and . . . later comes to *see* inquisitors and witches, where there are none." By contrast "hallucinations from without" were normally the result of either physical illness or injury, and he gave other examples such as "defects in the eye, or in the optic nerve" causing strange visions, or an ear infection producing buzzing sounds that "develop into hallucinatory voices."[70]

The subtleties of hallucination theory were essential when critically evaluating personal testimonies from people professing to have observed genuine spirit or psychic phenomena. With a strong grasp of the different ways in which the mind could be tricked into believing that an unusual vision or sound was real, researchers could more easily identify credible witnesses. A man seeing wraiths while his "brain was drugged with alcohol" was an unreliable witness, and so was "a man whose eyes are so vicious as to habitually give him false information."[71] Reports by witnesses where there was clearly no evidence to suggest that the percipient was suffering from any type of hallucination were needed. With Lang's detailed set of criteria, his readers could confidently evaluate the observations of witnesses using the descriptions he provided.

The fourth SPR case that Lang critically evaluated occurred in the south of Europe. "Mr. Harry" and his daughters repeatedly encountered the figure of a "white female" in various locations around their home. Mr. Harry was skeptical of spiritualism and possessed a "sturdy common-sense," apparently making him a reliable witness. However, much like the boy in case three, the ghost of the woman was usually seen in the bedroom when observers were in semi-consciousness. Lang recounted that on one occasion the figure of the white woman "lifted up the mosquito curtains and stared at Mr. Harry."[72] There was reason to believe that Mr. Harry was misinterpreting the situation, and that the visions of the white woman were the result of dream-induced hallucinations. Yet, as Lang contended, "A hallucination, we need hardly say, would not raise the mosquito curtains."[73] That sort of physical phenomena, mixed with the appearance of a ghost, complicated matters and suggested to Lang that something else was at work.

In the fifth case, we again have multiple reports by family members and servants of seeing a female apparition moving about a room in a house in Sussex. Lang stated, "One servant, who slept in the room in hopes of a private view, saw 'a ball of light with a sort of halo round it.' Again, in a very pretty story, the man who looked after an orphan asylum saw a column of light above the bed of one of the children. Next morning the little boy declared that his mother had come to visit him, probably in a dream."[74] In both cases, a skeptic could assert that the visions were the product of dream-induced hallucinations, but Lang provided an alternative explanation. Building on the work of the psychical researchers Podmore and Myers, he believed that some haunted houses were the product of telepathic thought transference. These extraordinary phenomena were the result of a spiritual residue produced by the memories of someone who had occupied the space previously. These memories allegedly lingered even after a person's death.[75]

Lang discussed one case where both Podmore and Myers provided alternative (but corresponding) explanations for an alleged haunting. They both agreed that some sort of telepathic force caused the unusual phenomena that were witnessed, but they disagreed on who was responsible for producing these disturbances.[76] The story began in 1886 when a woman named Miss Morris was deeply troubled by regular encounters with a supposed ghostly apparition in her home. After much distress, she left the property, never to return. In November 1887, "Mrs. G." took over the house, only to experience similar extraordinary phenomena. She heard "sobs, and moans, and heavy thumps, and noises of weighty objects thrown about." Seeking support, she summoned the help of barristers, clergymen, and even the police. All of them observed startling displays, but were unable to resolve the problem. An Anglican minister tried to exorcise the alleged entity, but the attempt was ineffective. The concurrent testimony from various observers of high moral and social standing gave much weight to the ladies' claims that the home was haunted. However, the ineffectiveness of the exorcism suggested that there was no actual spirit present, but something else.[77]

Podmore believed that the apparition and strange noises that Mrs. G witnessed were actually telepathic memories created by Miss Morris either from her time living in the house or whenever she reflected on her negative experiences there. Lang explained that "the later experiences may have been started by thought transference from Miss Morris, whose thoughts, no doubt, occasionally turned to the house in which she had suffered so much agitation and alarm."[78] These telepathic memories then manifested themselves as visual and auditory disturbances, which Mrs. G witnessed and heard. Myers, by contrast, argued that this explanation made no sense. If Miss Morris was responsible for creating the phenomena that was disturbing Mrs. G, what was causing the phenomena haunting her? Myers asserted that all the disturbances were produced by the "dreams of the dead." Apparently, a former occupant had committed suicide by hanging herself "with a skipping rope."[79] Her tormented memories lingered in the home even after her passing, and Miss Morris and Mrs. G were picking up on this telepathic thought transference.[80]

Lang's assessment of the SPR's five case studies served as a baseline to set up his critical reassessment of several famous ghost stories that were, for the most part, debunked by previous investigators. The problem, according to Lang, was that none of these researchers applied telepathic theories to their studies. If they had investigated the cases using the "anthropological method," combined with his revisionist version of animism, which included telepathy, it was possible to show that the recurrent appearance of similar sorts of extraordinary phenomena was the product of not simply credulity, superstition, and delusion, but telepathy. It probably was not spirits causing these unusual occurrences, but mediums using some inherent, unseen human faculty. To demonstrate the veracity of this theory, the evidence had to be tested, and once again it all hinged on the credibility of the witnesses. The comparative method would be essential for this process, but so too would Lang's revisionist animistic framework that included telepathy.

The Cock Lane Ghost

The Cock Lane ghost story is one of the most infamous accounts of a supposed fabricated haunting in modern British history. As Lang argued, it was "a proverb for impudent trickery, and stern exposure."[81] Although most skeptics tended to believe that it was a closed case, and that the original investigators provided ample evidence to prove that the alleged spirit was the product of normal human agency, Lang argued that the findings were actually uncertain. If the details of the Cock Lane ghost story were carefully reconsidered, a different conclusion might be formed. He argued that even at the time of the investigation "the very people who 'exposed' the ghost, were well aware that their explanation was worthless, and frankly admitted the fact."[82] Not only were the exposure and subsequent confession acquired through questionable means but the investigators were working with a false premise. A spirit was not causing the unusual disturbances. According to Lang, they were likely the result of unrecognized telepathic powers.

Similar types of disturbances were observed in countless other historical and ethnographic records. Lang wrote, "We can, indeed, study even the Cock Lane Ghost in the light of the Comparative, or Anthropological Method. We can ascertain that the occurrences which puzzled London in 1762, were puzzling heathen philosophers and Fathers of the Church 1400 years earlier. We can trace a chain . . . through the ages, and among races in every grade of civilization."[83] Lang devoted nearly an entire chapter in his book to an historical reconstruction of the affair, basing much of his analysis on reports from periodicals, in journals such as *Gentleman's Magazine*, and a pamphlet published by the Irish writer Oliver Goldsmith (1728–1774) titled *The Mystery Revealed* (1762).[84] If Lang were to prove that the phenomena occurring at the apartment on Cock Lane were caused by telepathic forces, he had to reassess the evidence and show his readers how a comparative examination of the material shed new light on the story. The application of microhistorical practices to create a "thick description" is rather useful in this instance for interrogating Lang's reconstruction of the case.[85] Not only does it mirror his own analytical processes, but by following along in the details of the case, it is possible to dissect Lang's "visual epistemology" for scrutinizing the evidence.[86]

The Cock Lane ghost came to prominence in 1762 and attracted mass public attention, but the proceedings leading up to the affair began in the late 1750s.[87] In 1756 a usurer named William Kent married a young woman named Elizabeth Lynes, and the couple moved to Stoke Ferry in Norfolk. To supplement his money lending, Kent and his wife also owned a small inn. Unfortunately for Kent, it was a short-lived marriage, as soon after moving into their new home, Elizabeth died in childbirth. During the later stages of Elizabeth's pregnancy, her sister Frances (or Fanny, as she is usually called) came to stay with them. She cared for Kent and the house, while her sister rested. After Elizabeth's death, Fanny stayed on to care for Kent during his period of grief, and soon the two began a romantic relationship. Kent and Fanny intended to marry, but as Lang recounted, "canon law would have permitted the nuptials, if the wife [Elizabeth] had not born[e] a child which lived, though only for a few minutes." Not wanting to cause any more problems, Kent chose to leave Fanny, and in 1759 moved to London. However, Fanny wanted to be with Kent, and despite her family's disapproval, she moved to London later that year to be with him. Unmarried, the couple struggled to find lodgings. Eventually, they got an apartment on Cock Lane (see fig. 3.1) after meeting Richard Parsons, a lay clerk working at Saint Sepulchre's Church near the Holborn Viaduct in London.[88]

Soon after moving to the Cock Lane apartment (see fig. 3.2), Kent lent Parsons twelve guineas, and this exchange became an important detail in the case. When the so-called haunting gained widespread attention in the press,

Figure 3.1. Woodcut of the exterior to the building where William Kent and Frances Lynes resided, on Cock Lane near Smithfield Market, London. *Source*: Charles Mackay, *Memoirs of Extraordinary Popular Delusions*, 2nd ed. (London: Office of the National Illustrated Library, 1852), 2:228.

many believed that Parsons and his family fabricated the ghost as a way of getting out of his debt to Kent. Initially, though, the two men had an amicable relationship. It was not long into their residency that Fanny became pregnant. Kent had to attend a wedding back in Norfolk, and Fanny asked Parsons's

Figure 3.2. Woodcut of the apartment on Cock Lane where the haunting allegedly occurred. *Source*: Mackay, *Memoirs of Extraordinary Popular Delusions*, 2:230.

oldest daughter, Elizabeth Parsons, if she would room with her during her husband's absence. This was when the disturbances started. During several of the nights they roomed together, both of them were kept awake by strange scratching and rapping noises. Mrs. Parsons—whose forename is unknown—initially attributed the sounds to "the industry of a neighboring cobbler." However, when the scratching returned on a Sunday, when it was clear that the cobbler was not in his workshop, this explanation was rejected.[89]

Fanny and the Parsons family came to believe that an unsettled spirit was causing the noises, and when Kent returned home from his trip to the countryside, he was told about the disturbances. Fanny pleaded with him to find new lodgings, and in January 1760 they moved to a nearby apartment in Bartlet's Court in Clerkenwell, London. Fanny died soon afterward of smallpox, but the drama of the Cock Lane ghost story was only just unfolding. Lang explained that "the noises in Cock Lane had ceased for a year and a half after Fanny left the house, but they returned in force between 1761 and 1762."[90] So began a series of investigations into the cause of the phenomena. All in all, over thirty people were involved in the case, ranging from medical practitioners, barristers, and noblemen, to carpenters, municipal politicians, clergy, and a lexicographer.

Several early theories were proposed to try and explain the phenomena. Some were fairly basic: the noises were caused by rats living in the wall, or a "mischievous neighbor" playing tricks. However, Parsons quickly dispelled these theories when he "took down the wainscoting" in the room with the help of a carpenter named Richard James, and no evidence of either was found. Other theories were more extreme. One such theory posited that the ghost of Elizabeth Lynes was wreaking havoc on her sister Fanny for being in a romantic relationship with her former husband. Another theory proposed that Fanny was having premonitions of her own death. It was eventually resolved that some séances should be held to see if the supposed spirit would communicate its purpose.[91]

It was clear that the person most affected by the disturbances was Elizabeth Parsons. Lang wrote that the "child 'was always affected with tremblings and shiverings at the coming and going of the ghost.'"[92] Some suspected that the girl possessed latent mediumistic powers. Through séances led by Richard Parsons and his family and friends, it was supposedly discovered that the ghost was none other than Fanny, who was trying to communicate to Elizabeth Parsons that she had been poisoned by Kent. The ghost was soon renamed "Scratching Fanny."[93] However, this revelation coincided with another development: Kent sought to sue Richard Parsons for the remainder of the money he, Parsons, had borrowed. No doubt framing Kent for the purported murder of his former lover was an attempt to discredit him, and perhaps avoid repaying the money owed. Many involved in the case were aware of this story, and most people quickly dismissed the accusation. There was also ample evidence to show that Fanny had not died of poisoning. At the time of her death, both the physician, a Dr. Cooper, and the apothecary, James Jones, who were on hand, confirmed her death as caused by smallpox.[94]

The disturbances continued, however, and eventually a group of skeptical investigators led by the biographer, literary critic, and lexicographer Samuel Johnson (1709–1784) thoroughly examined the case to determine the causes of the phenomena. One of Johnson's coinvestigators was the Scottish physician George Macaulay (d. 1766), who suspected that Elizabeth

Parsons used some sort of ventriloquism or hidden object to produce the noises. Lang discussed how on one occasion Macaulay "held his hand on the child's stomach and chest while the noises were being produced," but did not detect any trickery. He even had the girl "bound and muffled" to see if the disturbances would cease.[95] This practice of binding a suspected medium to see if their powers were legitimate, Lang stated, "is practised by Eskimo Angakut, or sorcerers, as of old by mediums in ancient Greece and Egypt."[96] Such a comparison allowed Lang to link his analysis to anthropological discourses on shamanism.

The investigation came to a head in early February 1762, when Elizabeth Parsons was moved to the home of the Anglican minister Stephen Aldrich for further observation. If it were through Parsons that the disturbances were channeled, they would occur regardless of whether she were at the apartment on Cock Lane or somewhere else. Moreover, because she had never been to Aldrich's home before, it was easier to establish controls, and limit the possibility of either her or a conspirator manipulating the space in advance of the investigation.[97] After carefully checking her body for any hidden device, Parsons was put to bed and observed by a group including Johnson, Macaulay, and another Anglican minister, John Douglas (1721–1807), who later became the Bishop of Salisbury. There were also some ladies on hand. After waiting for "more than an hour" without any disturbances occurring, the men retired to a separate room, while the women watched the child sleep. Johnson and his team were soon called back when scratching noises were heard, although they ceased as the men re-entered the room. Finally, after a few more days of unsatisfying results, the investigators threatened Parsons that if no noises were produced by the following night, she and her father would be "committed to Newgate" for fabricating the whole affair.[98]

The next evening while preparing Parsons for bed, some of Aldrich's maids saw the child conceal a little wooden board in her nightdress. The investigators were quickly informed, but waited to reveal this knowledge to Parsons in order to see first if noises were suddenly heard. Lang recounted that "scratches were now produced," and afterward the investigators exposed Parsons as a cheat by uncovering the hidden wooden board in her clothing.[99] A woodcut reproduction of the object was printed in 1762, and widely circulated throughout the periodical press. It was accompanied by a diagram depicting the layout of the room on Cock Lane where the noises originated (see fig. 3.3). These images vividly portrayed for readers how careful Johnson

The true Portrait of the G H O S T.
Taken from the Li█, and In-graved by

S. S. P. Sexton.

PARSNS Inv.

Scratch'd by one, two, three
and one More.

Plan of the Room, and the GHOST's Reprefentations, with
References.

The chimney,

Here was the fluttering.

The Bed

The wainfcot here, and

The moft
Knocking
was here.

here was the Scratching.

N. B. None but true Believers can make out the identical figure of the Apparition in this Picture. Infidels fee it as a confufed affair, fignifying Nothing.

Figure 3.3. Woodcut of the wooden board that Elizabeth Parsons concealed in her
nightdress. The diagram is of the layout of the room on Cock Lane where the noises
originated. *Source*: [S. S. P. Sexton], "The True Portrait of the Ghost," in *The Beauties
of All Magazines Selected for 1762*, ed. George Alexander Stevens (London, T. Waller,
1762), 1:48.

and his coinvestigators had been in carrying out their examination. Although the evidence presented in these images seemed to favor a verdict of cheating, the materials were strategically designed so that the investigation appeared to be completely candid, giving the impression that readers could form their own opinions based on both the textual and visual information available. This method of presenting the evidence added much weight to the claim that the whole affair was an act of deception. Despite these clear manipulations, Johnson's attention to recording the details of his investigation greatly assisted Lang in his later reconstruction of the event, thereby allowing him to inspect closely the supposed facts of Johnson's findings.

At this point in his retelling of the Cock Lane ghost story Lang began to critically re-evaluate the proceedings of the investigation. He argued that the discovery of the wooden board, and Elizabeth Parsons's subsequent confession, were achieved under duress. It was only upon being threatened that her family would be prosecuted for fabricating the affair that she decided to fake the noises. Yet even at the time of the investigation, several observers noted that "the noises the girl had made . . . had not the least likeness to the former noises."[100] Lang believed that the original disturbances were genuine phenomena, and he suggested that they were likely produced telepathically through Parsons. However, under hostile conditions, she was unable to produce the sounds telepathically, forcing her to take riskier measures by cheating. This would explain why the later fabricated sounds were different from the original ones. The investigation was therefore compromised because of the aggressive tactics used by the investigators. There were also class and gender imbalances to consider as well. Parsons was only a young girl from a disadvantaged background, and the investigators were predominantly older men from the upper classes of British society. It was easy for them to use their standings to pressurize and discredit her.

If Parsons had been given a fairer hearing, the conclusions might have been different. Comparing the Cock Lane ghost story to that of the early experiences of the Fox sisters in Rochester, New York, Lang argued that Parsons might have been the progenitor of the modern spiritualist movement, had she received a "more favorable environment" in which to present her mediumistic abilities.[101] Lang's critical reassessment of the circumstances surrounding the investigation of the Cock Lane ghost was therefore a significant attempt to challenge the mainstream narrative, which not only marginalized the supernormal elements of the story but denigrated them as little more than "impudent trickery."[102] His re-evaluation of the case was a clear attempt to empower belief in the potential reality of psychic forces by giving believers more agency in the fight against hardline rationalism and disbelief, which conflicted with Lang's more romantic and enchanted view of the world.

Employing the "anthropological test of evidence," Lang compared the Cock Lane haunting to a similar case at a farmhouse in Derrygonnelly, near Enniskillen in County Fermanagh, Northern Ireland. Much like the noises disturbing the apartment on Cock Lane, the unusual phenomena occurring at the house in Derrygonnelly took the form of scratching sounds. However, unlike the Cock Lane affair, where the investigators were ardent skeptics seeking to dismiss the occurrences through any conceivable means, the "Demon of Derrygonnelly," as it came to be known, received a far more balanced inquiry. The main investigator of the case was the physicist and experienced psychical researcher William Fletcher Barrett. Like many of his colleagues at the SPR, Barrett supported the possibility of telepathy being real.[103]

The investigation came about after Barrett and his friend, the geologist and archaeologist Thomas Plunkett, received thirty pounds from the British Association at Glasgow to explore some local limestone caves in Lough Erne, not too far from the farmhouse in Derrygonnelly. After a long day of surveying, Barrett and Plunkett visited the alleged haunted site. During the investigation, Barrett applied many of the same investigative techniques that Johnson and his colleagues had used in the Cock Lane inquiry. However, because he was more open to the possibility that some genuine unseen force might be responsible for the noises, the outcome of the inquiry was different. Similar to the Cock Lane investigation, the farmer's oldest daughter, Maggie, was most affected by the disturbances, and Barrett made the decision to monitor her and the other children while they slept.[104] Lang reported, "A soft pattering noise was soon heard, then raps, from all parts of the room, then scratchings, as in Cock Lane. When Mr. Barrett, his friend, and the farmer entered with a candle, the sounds ceased, but began again 'as if growing accustomed to the presences of the light.' The hands and feet of the young people were watched, but nothing was detected, while the raps were going on everywhere around, on the chairs, on the quilt, and on the big four-post wooden bedsteads where they were lying."[105] No deception was observed, and Barrett seemed satisfied with the conclusion that the disturbances were real. Lang wrote, "Mr. Barrett . . . could detect no method of imposture."[106] Much like Cock Lane, there was also a recent death affecting the family. The farmer's wife, mother to the children, had recently died, and the family were still grieving. The emergence of the noises happened just after the death, and Barrett believed that some sort of telepathic process could be causing the disturbances. Moreover, the farmer and his family gained little from forging the phenomena. Barrett stated that the household "reaped no profit by the affair, but rather trouble, annoyance and the expense of hospitality to strange visitors."[107]

Using a comparative approach, informed by his revisionist, animistic framework, Lang wanted to show that similar cases to the Cock Lane ghost story were recorded in countless historical and ethnographic sources, but when given a fairer assessment the outcome was very different. Using the example of Barrett's investigation of the Demon of Derrygonnelly, he argued that telepathy, or some other unseen human faculty, might be responsible for the extraordinary phenomena haunting the farmhouse. If similar treatment were given to all investigations of haunted homes, many cases might be confirmed as legitimate, and not the result of mere trickery or delusion. Lang was not trying to negate these possibilities, but he did believe that skeptics were too quick to dismiss unusual occurrences. More earnestness was needed before a conclusion could be reached as to the reality of the phenomena witnessed. Credible observers had to lead the investigations, and not allow their inherent biases affect the interpretations of their findings.

The Cideville Poltergeist

One of the last high-profile witchcraft trials of modern Europe occurred during the winter of 1851. In the small farming community of Cideville in Normandy, France, a shepherd named Felix Thorel stood accused of using his alleged magical powers to torment a local priest named Jean Tinel, and his young students Gustave Lemonier and Clément Bunel. In this trial, Thorel was suing Tinel for defamation of character. Thorel flaunted his alleged powers, but he was upset because he claimed that he was not responsible for the disturbances that occurred at Tinel's home. He argued that Tinel's accusations damaged his local reputation, and resulted in him losing his job. The case was well documented, with a rich surviving record. Forty-two witnesses were heard at the trial, and their concurrent testimonies added substance to the genuineness of the affair. Lang believed that the Cideville poltergeist, as it came to be known, was an important episode in the history of psychical research, offering compelling evidence in favor of the reality of telepathy and psychic phenomena. By applying the "anthropological test of evidence" to the case, it was also possible to show that the same types of disturbances occurring in Cideville were likely responsible for similar phenomena documented in other cases.

The story of the Cideville poltergeist was popularized among English-speaking audiences by Robert Dale Owen, who published an account of the trial in *Footfalls on the Boundary of Another World*. His version of the story was based on a pamphlet written by one of the chief investigators of the case, the marquis of Gommerville, Charles-Jules de Mirville (1802–1873).[108] However, Lang was highly critical of Owen's account because it disconnected the story from similar cases of witchcraft, such as the Salem witch trials between 1692 and 1693. Lang wrote, "Mr. Owen, by accident or design, omitted almost all the essential particulars, everything which connects the affair with such transactions as the witch epidemic at Salem, and the trials for sorcery before and during the Restoration."[109] Without this comparison, Lang believed, Owen devalued the significance of the material. If handled correctly, however, using Lang's revisionist animistic framework, the Cideville poltergeist story could be used to show that all cases of witchcraft— both historically and globally—were not the result of trickery, superstition, or delusion, but of unrecognized telepathic processes. Its recurrence in all cultures demonstrated that it was a universal human phenomenon, and Lang argued, "A more astonishing example of survival cannot be imagined."[110] It was, therefore, further evidence that could potentially legitimize telepathy.

The story began in March of 1849, when Tinel visited the farm of a sick peasant suffering from an unknown malady. Tinel believed that the illness was caused by some magical curse inflicted on the peasant by an elderly shepherd named "Old G," who supplemented his income by practicing folk medicine and sorcery. Tinel suggested to the peasant that he banish Old G from his household, and bar him from using his so-called powers on him again. The elderly shepherd was hiding nearby, and overheard the conversation. Angered by what had happened he allegedly stated: "Why does he [Tinel] meddle in my business, I shall meddle in his; he has pupils in his house, we'll see how long he keeps them."[111] A few days later, Old G was caught practicing unauthorized medicine at another home, and was imprisoned for several months. He believed that Tinel played a key role in his conviction, and seeking revenge, he enlisted the help of his friend, Thorel, who was also known locally for practicing folk magic.[112]

For a curse to be laid on Tinel's students, it was alleged that they had to be touched by the sorcerer, and in late November of 1850 Thorel encountered Tinel's students Lemonier and Bunel at a local wood sale. He cursed them with his so-called magical touch, and later boasted of his success to some local villagers. The following evening, strange occurrences began, and Tinel and his students heard some "light blows of a hammer" while studying in the priest's drawing room. When Tinel shouted "Plus fort," the noises became louder. Soon, more unsettling phenomena occurred: furniture moved around the house untouched, and a knife allegedly flew around a room uncontrollably.[113] Eventually, Lemonier was haunted by a phantom with a resemblance to Thorel, though the boy did not yet know of this similarity. One day Thorel visited Tinel's home, and, as Lang recounted, the boy exclaimed: "There is the man who follows me." Tinel then forced Thorel to beg Lemonier's pardon, and afterward told Thorel's employer to relieve him of his duties. Out of work and disgraced, Thorel sued the priest for libel.[114]

At this point in the narrative Lang shifted his focus to the reliability of the boys' testimonies. Not only were they priests in training, which supposedly raised the trustworthiness of their observations because of their so-called high moral characters, but they had also sworn oaths during the trial, binding them legally and morally to providing honest reports. One piece of testimony that became particularly important for Lang was a description given by Lemonier in which he claimed that the phantom haunting him was struck by a nail on his cheek, and when it was revealed that Thorel had a similar cut on his face, the boy's claim strengthened the allegation that Thorel was responsible for the phenomena. Even if he was not physically present, there was a long history of witchcraft cases where injuries caused to an apparition affected the conjurer. Lang wrote, "This is in accordance with good precedents in witchcraft. . . . At the trial of Bridget Bishop, in the court of Oyer and Terminer, held at Salem, June 2, 1692, there was testimony brought in that a man striking once at the place where a bewitched person said the *shape* of Mrs. Bishop stood, the bewitched cried out, *that he had tore her coat*, in the place then particularly specified, and Bishop's coat was found to be torn in that very place."[115]

Lang strategically applied the comparative method to situate the Cideville poltergeist story within the broader context of witchcraft. Using well-known sources such as Cotton Mather's *Wonders of the Invisible World* (1693), Lang was attempting to strengthen the legitimacy of Lemonier's report.[116] The message was simple: these extraordinary phenomena were created through human agency, and if you harmed an apparition that was produced through some sort of unseen human faculty, you harmed the supposed conjurer, too, thus exposing its true source and confirming that it was not the product of spiritual energies, superstition, or delusion. The concurrent testimonies in all cases of witchcraft—from historical and ethnographic records—suggested that there was some real underlying force causing these occurrences. Through the "anthropological test of evidence," Lang believed, it was possible to affirm that all cases of witchcraft were actually instances of unregistered telepathy. Beneath the superstition and religious fervor, these cases documented examples of genuine mediumship, and for Lang, figures such as Bishop in the late 1600s, or Thorel in the 1850s, were no different from Daniel Dunglas Home, William Stainton Moses, or Indigenous spiritual figures such as Inuit angakut or Aboriginal Australian kurdaitcha.[117]

Putting aside these other cases, there was still ample evidence to show that Thorel possessed extraordinary powers. Lang argued that forty-two witnesses were heard at the trial, ranging from local landowners, surveyors, and municipal officials to clerics from neighboring parishes, farmers, and psychical researchers such as the marquis de Mirville. All of them attested to witnessing strange phenomena at Tinel's home, or of knowing other inexplicable feats that Thorel had produced on different occasions.[118] For example, when the mayor of Cideville, Adolphe Cheval, took to the stand, he recalled one story where the shepherd used his supposed powers to torment a fellow sheepherder. Every time Thorel struck his caravan, his associate was thrown to the ground by some unseen force.[119] Thus, he had a long history of misusing his so-called magical abilities to pester others.

Thorel's most impressive feat—at least according to witnesses at the trial—was the ability to project hallucinations onto other people. Lang interpreted this act as a type of telepathic thought transference, and compared it to similar cases from the SPR records. He wrote, "in the *Proceedings of the Society for Psychical Research* [there] is a long paper by Dr. Gibotteau, on his experiments with a hospital nurse called Berthe."[120] According to the report,

Berthe had the ability to make the doctor see frightening hallucinations even when they were not together. She could also make people stumble or fall without touching them. In both cases these sorts of phenomena were similar to the kind Thorel produced in Cideville, and Lang argued that they were typical of a sort of "rural sorcery" that was common in the French countryside. If one accepted Gibotteau's testimony as trustworthy, and that the nurse genuinely performed these feats, then by extension the reports from the Cideville poltergeist trial should also be taken seriously. Given that Gibotteau was a medical expert with comprehensive training in psychological research, there was no reason to doubt the integrity of his account.[121]

After listening to the various testimonies, the judge presiding over the trial of Thorel versus Tinel dismissed the case as a frivolous lawsuit. Thorel could not possibly sue Tinel for slander if he openly admitted to possessing magical powers. The judge ruled that Thorel had to pay for the court costs, and he absolved Tinel of any wrongdoing.[122] Yet because the court did not attempt to reject the disturbances as fake, it opened up the possibility of using the evidence to support the existence of supernaturalism. Lang's analysis of the Cideville poltergeist story did exactly that: it used anthropological methods to demonstrate that Thorel, and similar figures such as Bridget Bishop and Berthe, were verifiable examples of telepaths.

Fundamentally, the crux of the Cideville poltergeist story rested on the testimonies of Lemonier and Bunel, who provided the most detailed accounts of the disturbances. Skeptics reading the facts of the case could, of course, argue that there were no genuine telepathic or magical powers on display. Thorel likely sneaked into Tinel's house, and played some nasty tricks on the boys. Their later testimonies at the trial were nothing more than exaggerations of their experiences. Or, given that the boys were always around when the phenomena were present, it might also be possible that they were the true culprits responsible for the disturbances. It is reasonable to suspect that they concocted all of the alleged occurrences in an effort to seek revenge on their master's rival. Nevertheless, Thorel's earlier boasting, in which he took credit for the acts, combined with similar examples from other cases of witchcraft, was enough for Lang to believe that the phenomena were authentic and explainable through telepathic processes. However, for the stubborn skeptical reader who was still unconvinced by his analysis, Lang had another case study to examine, one that took seriously the importance of scientific processes for assessing alleged spirit and psychic phenomena: experiments with table turning. Here again, we see the application of his revised animistic framework, which featured the possibility of real telepathy.

Agénor Étienne, Comte de Gasparin, and His Experiments with Table-Turning

Victorian naturalists were understandably skeptical of investigations into the reality of spirit and psychic phenomena because of its links to supposed "primitive superstition." For Lang, though, his investigations were not about questioning people's beliefs in spiritualism but about determining if the occurrences were real and based on tangible facts. He argued, "Some men of science denounce all investigation of the abnormal phenomena of which history and rumour are so full, because the research may bring back distasteful beliefs, and revive the 'ancestral tendency' to superstition. Yet the question is not whether the results of research may be dangerous, but whether the phenomena occur."[123] These preoccupations firmly rooted spirit and psychic investigations in epistemic debates over the veracity of animism, and to meet this challenge, Lang argued that it was the duty of scientists to take the study of extraordinary phenomena seriously and assess its legitimacy. A good starting point was table turning, which was one of the first phenomena to be observed during the early days of the modern spiritualist movement. As with the Cock Lane ghost and the Cideville poltergeist, Lang carefully reconstructed the history of Agénor Étienne, Comte de Gasparin's experiments with table turning.

Since the emergence of the Fox sisters in the late 1840s, table turning had been a regular feature of most modern séances. Already in the 1850s, investigators were studying the movements and sounds of so-called possessed tables to determine if these bizarre displays were the product of trickery and delusion, or some other unseen force. News of this strange phenomenon reached Germany in 1853, and Lang remarked that "all [of] Bremen took to experiments in turning tables. The practice spread like a new disease, even men of science and academicians were puzzled."[124] Hearing of these reports from Germany, the French writer and statesman de Gasparin began investigating spiritualism. His experiments with table turning represented one of the first prolonged studies to take seriously the validity of these occurrences, and for Lang it provided early scientific evidence in support of telepathy.

Lang used de Gasparin's results to support his revisionist model of animism, and he framed the Frenchman as a philosophically minded observer and skilled experimenter. He made a strong case for de Gasparin's honesty and high moral integrity, stating that "M. de Gasparin, in the early Empire, was a Liberal, an anti-Radical, an opponent of negro slavery, a Christian, [and] an energetic honest man."[125] In 1853 de Gasparin conducted tests on table turning, which were eventually published in English as *Science vs. Modern Spiritualism* (1856).[126] His results were quickly contested, however, and one of his early adversaries was the French chemist and physician Louis Figuier (1819–1894). According to Figuier, de Gasparin was never able to reproduce his results in front of the French scientific community, making his observations untrustworthy. Figuier's critiques were later published as part of *Les Mystères de la Science* (1880).[127] Lang was unconvinced by these criticisms, and he was a staunch defender of de Gasparin's findings. He included the details of de Gasparin's experiments in *Cock Lane and Common Sense*, championing the Frenchman's research practices, and crediting him as a trustworthy observer. The reconstruction of de Gasparin's experiments was also an exercise in legitimizing the veracity of the findings. It was a classic example of what Steven Shapin and Simon Schaffer have termed "virtual witnessing," a process by which readers could mentally recreate experimental tests, as if they too, had directly participated in them. This sort of thought experimentation had been a staple of modern science since the seventeenth century, and was essential to Lang's visual epistemology in *Cock Lane and Common Sense*.[128]

One of the keys to establishing de Gasparin's authority as a psychical researcher was to show that he was not a spiritualist. He actively tried to dismiss the idea that spirits were responsible for moving the tables. He was equally critical of the physiological explanations that were circulating at the time regarding "unconscious muscular movement."[129] This balance of critical perspectives substantiated his observations. Lang wrote, "His purpose was to demonstrate that tables turn, that the phenomena is purely physical, that it cannot be explained by the mechanical action of the muscles, nor by that of 'spirits.'"[130] Some unseen psychic force of human origin moved them. However, to prove this point, he had to thoroughly demonstrate that the more naturalistic explanations were wrong. Typically, de Gasparin would get a group of eight to ten people to position themselves around a table and gently rest their fingertips on its surface (see fig. 3.4). On request, the table

Figure 3.4. An unknown force tilts a table while a group of sitters feel its movements by gently resting their hands on it. The woodcut most likely represents one of de Gasparin's experiments with table turning during 1853. The image is included in Figuier's examination of de Gasparin's experiments from 1853. See Louis Figuier, *Les Mystères de la Science* (Paris: Librairie Illustrée, 1880), 2:576.

would eventually start to move. Other experiments checked for pressure, and Lang wrote, "Two persons for whom the table would *not* move laid their hands on it firmly and flatly. Two others (for whom it danced) just touched the hands of the former pair. Any pressure or push from the upper hands would be felt, of course, by the under hands. No pressure was felt, yet the table began to rotate."[131]

Much of this action could still be attributed to unconscious muscular movement, but de Gasparin circled around the issue by recording instances in which there was no physical contact with the table by the séance sitters, and yet it still moved. Lang reported, "He therefore aimed at producing movement *without* contact. In his early experiments the table was first set agoing by contact; all hands were then lifted at a signal, to half an inch above the table, and still the table revolved."[132] It was possible that one of the sitters inadvertently touched the table to keep it moving, but Lang contended that other reliable witnesses were present to assist de Gasparin in ensuring that no physical contact was made after the initial thrust. One of them was the professor of natural history and physics at the University of Geneva Marc Thury (1822–1905), who published corresponding findings in 1855.[133] Lang wrote, "The observer, Mr. Thury, saw the daylight between their hands and the table, which revolved four or five times. To make assurance doubly sure, a thin coating of flour was scattered over the whole table, and still it moved, while the flour was unmarked."[134] Thury's report added further credibility to the experiments, assuring readers that everything was done completely aboveboard and with controls in place.

If no physical contact was made with the table, its movements could not be caused by a push or jolt. Therefore, de Gasparin argued that his experiments with table-turning dispelled Michael Faraday's theory of "unconscious pressure," and William Benjamin Carpenter's arguments about "ideomotor responses."[135] He also challenged Carpenter's view that participants at séances were probably under some sort of hypnotic trance. There was no evidence to suggest that his sitters were disconnected from the proceedings, and all of them seemed quite alert and responsive to any instructions or requests given to them. Lang wrote, "M. de Gasparin averred that no single example of trance, rigidity, loss of ordinary consciousness, or morbid symptoms, had ever occurred in his experiments."[136] The data clearly indicated that some other force was in operation, and de Gasparin asserted that "There is, in man's constitution, a 'fluid' which can be concentrated by his will" and cause objects such as tables to move without contact.[137] It was essentially an early theory of telepathy. For Lang, the case offered compelling evidence to support his revisionist model of animism, because all of the standard explanations in Tylor's version—superstition, trickery and delusion—were dismissed by de Gasparin, and what remained (at least in Lang's view) was an obvious example of telepathy.

Carpenter revisited the topic of table turning in the 1870s, but there was no mention of de Gasparin's experiments in his work.[138] Lang argued that this devalued Carpenter's defense of his and Faraday's original ideas. He wrote,

> Dr. Carpenter . . . published a work on *Mesmerism, Spiritualism*, etc. Perhaps the unscientific reader supposes that Dr. Carpenter replied to the arguments of M. de Gasparin? This would have been sportsmanlike, but no, Dr. Carpenter firmly ignored them! He devoted three pages to table-turning (pp. 96, 97, 98). He exhibited Mr. Faraday's little machine for detecting muscular pressure, a machine which would also detect pressure which is *not* muscular. He explained answers given by tilts, answers not consciously known to the operators, as the results of unconscious cerebration. . . . But not one word did Dr. Carpenter say to a popular audience at the London Institution about M. de Gasparin's assertion, and the assertion of M. de Gasparin's witnesses, that motion had been observed without any contact at all. He might, if he pleased, have alleged that M. de Gasparin and the others fabled; or that they were self-hypnotised, or were cheated, but he absolutely ignored the evidence altogether.[139]

According to Lang, Carpenter could have at least acknowledged de Gasparin's work before trying to dismiss it through the usual means—that is, as being the result of superstition, delusion, or credulity. However, by completely ignoring it, Carpenter was actually giving credibility to de Gasparin's findings. The omission suggested that Carpenter could not disprove the Frenchman's results, and the absence of the information opened the door to the possibility that telepathy was provable. This was not the first time that Carpenter had purposefully ignored work that contradicted his skeptical views on spiritualism and mediumship. Alfred Russel Wallace also argued that Carpenter was somewhat dishonest in his critique of spirit and psychic investigations by ignoring so-called positive evidence rather than critically evaluating it.[140]

With the evidence considered, Lang shifted his argument back to the question of belief. He argued that the real problem was that skeptics were working under a false assumption. The conservative view was that so-called savages attributed extraordinary phenomena to external entities acting on the living. He wrote, "Epilepsy, convulsions, hysterical diseases are startling affairs. . . . It was natural that savages and the ignorant should attribute them to diabolical possession, and then look out for, and invent, manifestations of the diabolical energy outside the body of the patient, say in movements of objects, knocks, and so forth."[141] However, what if these phenomena were human in origin, and it was through telepathic processes that these unusual occurrences were created? Lang believed that his critical reassessment of cases such as the Cock Lane ghost, the Cideville poltergeist, and de Gasparin's experiments with table turning provided compelling evidence in support of telepathy, or some other unseen human faculty. Tylorian animism was left wanting, and for its proponents to uphold the theory they had to acknowledge that not all extraordinary phenomena were the result of credulity, delusion, or superstition. Accepting Lang's revisionist model, which included telepathy, was a viable solution.

Rejecting the Revisionist Model

Tylor's supporters were not so easily swayed by Lang's arguments. One of Lang's chief opponents was the popular science writer, anthropologist, and folklorist Edward Clodd. In his presidential address to the Folklore Society in 1895, Clodd dismissed psychical research as a pseudoscience, practiced by credulous fools who were easily deceived by inadequate evidence and overly susceptible to superstitious nonsense. Researchers such as Lang, Myers, Podmore, and others were disseminating a form of animistic belief disguised in the language of modern science.[142] Rational thinkers, though, should not be so easily persuaded by such ridiculous rhetoric. Even if it could be proven that telepathy was real, Clodd remarked that "much could be said about the absence of any contribution of the slightest value" this knowledge would provide for humankind.[143] Lang was quick to defend psychical research, and there were a couple more exchanges between him and Clodd in the pages of the journal *Folklore*. Both men made jibes at one another, but no consensus on the matter was achieved.[144] Animism and spiritualism were both questioned, but there was much uncertainty over whether a revisionist model that included telepathy was convincing enough to displace either theory. For the time being the causes of spirit and psychic phenomena remained unresolved.

Although Lang had been working on this subject for many years, *Cock Lane and Common Sense* represented the maturation of his thoughts. There was an overwhelming number of testimonies by witnesses professing to have observed genuine spirit and psychic phenomena, but if there was any weight to these claims then they had to be thoroughly scrutinized and still hold true. Similar cases had to be compared for consistency. Through an examination of three case studies—the Cock Lane ghost, the Cideville poltergeist, and de Gasparin's experiments with table turning—Lang constructed his revisionist animistic framework in support of telepathy. Hallucinations, trickery, and superstition were all part of Lang's investigation, but he still argued that there were plenty of examples where these explanations failed to account for the occurrences witnessed by credible observers. There had to be another option, and telepathy seemed in his view to be the most viable. He therefore became a leading proponent and voice for belief in psychic forces. His ideas were not left unchallenged, though, and he had his critics, especially Clodd. The debate over animism and spiritualism raged on, and Clodd would become a leading challenger and disbeliever of spiritualists and mediums in the decades ahead.

Edward Clodd

The Disbeliever

Edward Clodd is one of the more renowned science popularizers of the late Victorian age.[1] His twin interests of evolutionary theory and cultural anthropology dominated much of his writings. After reading *Primitive Culture* in 1871, he became an ardent follower of Tylorian anthropology, which culminated in the publications of his popular works *The Childhood of the World* (1873) and *The Childhood of Religions* (1875). Animism was a principal component of his theoretical model, and he remained intellectually committed to the theory right up until his death in 1930. Because of his preoccupation with the anthropology of religion, Clodd's attention was eventually turned toward the growing presence of spiritualists in Victorian Britain. His response to the movement's rise was similar to that of Edward Burnett Tylor, in that he viewed the belief as a survival and revival of so-called primitive superstitious thought. However, unlike Tylor, or even Andrew Lang or Alfred Russel Wallace, for that matter, Clodd was far more hardline in his rejection of the spirit hypothesis and telepathy. Regardless of any piece of evidence one could put before him, Clodd maintained an unwavering disbelief in spirit and psychic phenomena, and from the 1890s onward spent much of his energy refuting the claims of spiritualists and telepathists.[2] His engagement with modern spiritualism and psychical research, therefore, represents an important example of the disbeliever's perspective, and is a considerable departure from the positions of Wallace, Tylor, and Lang.

Clodd's disdain for spiritualism came to a boiling point during the horrific years of the First World War, when Britons experienced a level of mass death not seen since the pandemic of the bubonic plague in the fourteenth century. Many families were desperately clinging to the memories of their lost loved ones, and hoped that spiritualism might provide some solace. If spirit phenomena were genuine, then heartbroken mourners could continue to communicate with the deceased in the afterlife, lessening their sorrow.[3] The movement, therefore, experienced a sort of renaissance in the years surrounding the Great War. Clodd believed that mediums were capitalizing on the desperation of grieving people, and that these activities were deplorable. This is one reason why he went on the offensive, becoming relentless in his attack on spiritualistic and telepathic beliefs in the opening decades of the twentieth century. In 1917, he published *The Question: A Brief History and Examination of Modern Spiritualism*, which was his most detailed refutation of the modern spiritualist movement.[4] This was followed by a couple of high-profile lectures on occultism that he delivered at the Royal Institution (RI) in 1921, which offered a further set of damning critiques of spiritualism and psychical research.[5] Clodd believed that "modern psychism is but savage animism writ large."[6] Its foundation was completely unsteady, and for the spirit hypothesis to hold firm, the evidence supporting it had to be weighed carefully. Because the main evidence to support the existence of spirits and psychic forces came from the personal accounts of various types of observers, the crux of Clodd's analysis was devoted to a critical reading of witness testimonies. In the preface to *The Question* he wrote, "The subject of this book is not a history of the origin of the belief in immortality, but an examination of the evidence on which those who call themselves Spiritualists base that belief."[7] The comparative method provided the practical scaffolding of his study, and animism supplied its theoretical foundation.

For Clodd, Tylorian animism held the key to unraveling belief in spirits and psychics because all unusual occurrences could be explicated through causes such as delusion, illness, and superstition. Every "confirmed" case of spirits or psychic forces was merely a misinterpretation of events. It was up to vigilant investigators to identify the reasons for these errors of perception.

The key to refuting the spirit hypothesis was to create reasonable doubt. If an investigator could suggest rational alternative explanations for the causes that created extraordinary phenomena, the accounts could be dismissed as unreliable. Any inconsistencies, ambiguities, or fallacies could be used to delegitimize a medium's alleged powers. "Virtue epistemics," to borrow the term from Lorraine Daston and Peter Galison, was a key part of Clodd's analysis, and he focused considerable attention on the character and integrity of different kinds of witnesses.[8] Equally, if it could be shown that investigators did not have full control over the setting in which a medium performed psychic feats, these accounts could also be dismissed as untrustworthy. If one type of psychic feat was routinely exposed as fake, then by extension all instances of this kind of occurrence should be rejected as fake. It did not matter if every single instance of spiritualism or telepathy were exposed as fraudulent; all that mattered was that enough doubt was created to question the whole enterprise. Any gaps in the history of exposure did not in itself undermine the integrity of the skeptic's argument. There was sufficient evidence to support the case for rejecting the spirit hypothesis as nothing more than a survival and revival of primitive belief.

Clodd remains a relatively elusive figure within the secondary literature on the history of British anthropology, and much of this has to do with the fact that he was not an academic or scientific researcher, but a banker and popular writer. Bernard Lightman and Douglas Lorimer have written the most detailed accounts of Clodd's career. Lightman has focused on the important role of Clodd as a disseminator of evolutionary theory to nonspecialist audiences (including youths and adults) during the nineteenth century.[9] By contrast, Lorimer has shown the impact and legacy of Clodd's popular writings on conceptions of race and racism from the late Victorian period onward.[10] Within the historiography on Victorian spiritualism and psychical research, reference to Clodd's work is even sparser, with only fleeting mentions of him.[11] He has yet to receive any serious attention. This is rather significant given how prominent he was during the late Victorian era. As his biographer Joseph McCabe (1867–1955) recalled, "Few of the outstanding men of the late Victorian period had quite so interesting a development," and Clodd became "the intimate friend of an usually large group of the most distinguished representatives of letters, science, and art, a respected figure in the financial world, an anthropologist of note, [and] a writer whose works appear in a dozen languages."[12]

Most scholars today will probably identify Tylor as the chief Victorian anthropologist to propagate the theory of animism during the second half of the nineteenth century. However, during the late nineteenth century, many Victorians learned about Tylorian anthropology through the works of Clodd. He was a far more familiar name, and his books outsold any academic or scientific publication produced by Tylor, with some of Clodd's books selling in excess of twenty thousand copies to a broad Victorian readership.[13] As a bestselling author, his book on spiritualism would have reached a diverse audience as well. This raises an important issue. If it was predominantly through Clodd's popular anthropological works that Victorian Britons grappled with the consequences of animism, particularly its links to spiritualism, then a study of his work seems crucial.

The growth of spiritualism and telepathy in the opening decades of the twentieth century would have been deeply concerning to anthropologists forwarding a Tylorian model. This was a period of institutionalization for the young discipline, when researchers were vying for positions within universities.[14] If a core theoretical principle such as animism was undermined by something that seemed as preposterous as alleged spirit and psychic phenomena, anthropology's reputation as a credible scientific pursuit would be suspect. Clodd's staunch opposition to the veracity of the spirit hypothesis had one primary aim—to strike a final blow to extraordinary belief, and demonstrate rationally that there was no evidentiary basis to support the modern spiritualist movement or mediumship.

From Baptist Preacher in Training to Popular Anthropologist

Although Clodd would one day become a bestselling popular science writer, and champion of agnosticism and rationalism to a broad audience during the closing decades of the nineteenth century, his upbringing was the complete opposite, and deeply Christian.[15] The son of a Baptist brig captain, Clodd grew up in Aldeburgh, in Suffolk, where he attended the local grammar school, and received a highly religious education. His early schooling was filled with Christian literature by authors such as John Bunyan (1628–1688), Richard Baxter (1615–1691), and James Hervey (1714–1758). He recalled in his autobiography that "As for books, they were far to seek in any number. The Bible, [Bunyan's] *Pilgrim's Progress* and the *Holy War*, Baxter's *Saints' Everlasting Rest*, Hervey's *Meditations among the Tombs* and some odd missionary magazines, complete the serious list."[16] As for secular reading material, Clodd had limited access to it, and most "worldly books" were taboo among the local community. He was taught that the Bible provided the sole truth, and that he should fear the wrath of God.[17] It was his parents' hope that he would become a preacher, and his early teens were spent under the tutelage of the local Baptist minister. He recalled that his days were strict and disciplined, and that he was regularly assigned short essays or "little sermons" on "ambitious subjects [such] as the 'Abolition of Slavery' and the 'Character of Oliver Cromwell.'"[18]

His plans to serve the Baptist Church, however, changed after a visit to the Great Exhibition with his mother in 1851. Suddenly, Clodd's mind was opened up to a world of knowledge that had been hitherto unavailable to him in the Suffolk countryside. He wrote in his autobiography that "to a boy of eleven, whose farthest jaunts had been on holidays to relatives at Framlingham, this was to enter a wonderland which surpassed all that his mind could conceive. It made me secretly resolve, whatever might block the way, to get to London when I left school."[19] And leave school he did, when at the age of fifteen on a visit to London to see his aunt and uncle, Clodd seized the opportunity to find a job as an accountant's clerk, where he served his employer for six months without pay in order to gain work experience. He held a few more paid clerkships before eventually finding a job at the London Joint Stock Bank in 1862. As an ambitious young man, Clodd rose quickly within the institution, becoming the secretary in 1872, a position he held until his retirement in 1915.[20]

With steady income, and a home centered in London, Clodd took advantage of the many intellectual amenities that the city had to offer. During the 1860s, he further expanded his knowledge of multiple subjects, while studying at the Birkbeck Literary and Scientific Institution (f. 1823). In particular he "devoured books on science and history," two fields that would greatly influence his later anthropological writings on spiritualism. However, it was two publications from 1859 and 1860 that really changed Clodd's intellectual mindset: Charles Darwin's *Origin of Species*, and the chemists Robert Bunsen and Gustav Kirchhoff's paper on spectroscopy.[21] Both texts allowed him to question the biblical narrative of creation, and proposed naturalistic understandings of the origin of the life. Clodd's slow conversion toward scientific naturalism and agnosticism began henceforth.

Other major events during this period further shook Clodd's theism, and he recalled that "in June 1860, at the meeting of the British Association at Oxford, there was a memorable duel between Huxley and Bishop Wilberforce on the question of man's fundamental relationship to the great apes."[22] Clodd was utterly convinced by Huxley's arguments against special creation, and the evolutionary links between humans and animals. It seemed as though the religion of his youth was quickly fading.[23] Clodd was increasingly disposed to base his opinions on tangible scientific evidence, and not on abstract religious belief. This shift in trust from a priori knowledge to prima facie evidence formed the backbone of his later refutation of the modern spiritualist movement. Belief alone was not enough to sway Clodd to accepting the plausibility of the spirit hypothesis. Only concrete, reproducible data, collected by skilled investigators, could do that. If such data was unavailable, then personal testimony could be accepted only when the witnesses were thoroughly vetted for their trustworthiness.

Major works in biblical criticism from the 1860s also contributed to Clodd's anthropological model of religion. For instance, in 1860 the Oxonian classicist and theologian Benjamin Jowett published his famous essay "On the Interpretation of Scripture," which proposed an allegorical reading of the Bible. Clodd reminisced that "my waning belief in the Bible as in any sense a Revelation was shattered by reading Jowett's article."[24] This was followed a few years later by John Robert Seeley's anonymously published book, *Ecce Homo* (1865), which historically contextualized Jesus's moral character. Clodd recalled that "*Ecce Homo* sent a shudder through every sect in Christendom; through Nonconformists as well as [Anglican] Churchmen."[25] The result for Clodd, at least, was a dramatic reconsideration of his theistic upbringing. Religious texts were no longer the word of God to him, but culturally constructed works that were steeped in metaphorical meaning. It was then up to the critical reader to trace the origins of these beliefs through historical investigations. It was a methodological approach similar to that articulated by Tylor in his anthropological writings, and one that Clodd appropriated in his scathing attack on spiritualism.

It was also during the 1860s that Clodd immersed himself in the writings of scientific naturalists. He was particularly influenced by Thomas Huxley's *Man's Place in Nature* (1863) and *On Our Knowledge of the Causes of Organic Nature* (1863). The latter publication was originally delivered as a course of lectures at the Museum of Practical Geology to workingmen between 1860 and 1862.[26] As Clodd recounted, Huxley's core argument purported "that no barrier exists, either in body or mind, between man and animal, and that 'even the highest faculties of feeling and of intellect begin to germinate in lower forms of life.'"[27] Thus, the thesis of special creation was undermined by a naturalistic explanation for human evolution. However, of all the works to radically reorient Clodd's intellectual mindset, Tylor's *Primitive Culture* was arguably the most influential.[28] By reading this book, Clodd adopted the theory of animism as the best explanation for understanding the reasons for spiritual belief in the broadest sense. Combining his newfound knowledge of naturalistic evolution and biblical criticism, Clodd saw all religions as survivals of primitive superstitious thought. By the start of the 1880s he no longer identified as a theist, and adopted the agnosticism of the scientific naturalists.[29]

Huxley's and Tylor's respective works also introduced Clodd to the world of popular science writing, showing him the possibilities that the genre offered as a way of communicating knowledge to a broad nonspecialist audience.[30] He was particularly concerned by the potential damage that untruths about the origin of humans had on children, and he stated in his autobiography that "Satisfied, by study of these, and other books bearing on the subject, as to Man being both in body and soul no exception in the processes of evolution, and as to his history being one of advance from savagery to civilization, there followed concern as to what should be taught to my children. Were they to learn a mass of fiction, with the cost and pain of unlearning it afterwards, discovering that what I had taught them or allowed them to be taught was not the truth, the truth which alone can make us 'free'?"[31] The result of these anxieties was the publication of Clodd's first two books, *The Childhood of the World* followed two years later by *The Childhood of Religions*. Both texts were written for a young readership, using accessible language to tackle a difficult subject. The influence of both Tylor and Huxley were blended into a synergetic theoretical model. Clodd put forward a naturalistic evolutionary model based on Huxleyan principles to explain human physical history, and applied Tylorian animism to explain human cultural history.[32]

Clodd would continue to be one of the main proponents of animistic theory for the remainder of his life, and when spiritualism's popularity in Britain rose during the second half of the nineteenth century, he was one of the most vocal critics within anthropological circles. It was also during the last decade of the nineteenth century that Clodd became increasingly involved with the Rational Press Association (f. 1885), eventually becoming its chair between 1906 and 1913. He was also a council member of the Secular Education League in 1907. These activities further strengthened his commitment to materialism and scientific induction.[33] However, it was during his presidential address at the Folklore Society in 1895 that he delivered his first sustained attack on the modern spiritualist movement, and his primary foe was Lang, a revisionist of classic Tylorian animism.

Debating Spiritualism and Psychical Research at the Folklore Society

As we saw in the previous chapter, folkloric studies were a research field allied to anthropology during much of the nineteenth century, and many researchers, including Tylor, Lang, and Clodd, were active within both scholarly communities. Folklorists and anthropologists alike were interested in historically tracing the allegorical meaning of different practices and beliefs, using animism as a way of explaining how these customs, habits, traditions, rituals, and so forth survived into modern society. It should, therefore, not be surprising that Clodd's attack on spiritualism, which rested on his adherence to animistic theory, occurred at the Folklore Society in 1895. He would have found an attentive and sympathetic audience there. It was also a targeted attack on Lang, who had been championing the value of psychical research at society meetings. Clodd wanted to halt these discussions before the appeal of psychical research grew among the members.

Clodd's disdain for spiritualism was extended to psychical research because he believed that it was essentially the same thing, hiding under the pretense of an alleged science. He wrote in his presidential address to the Folklore Society (f. 1878) that the only difference between the Society for Psychical Research (SPR) and spiritualists was "the degree of certainty which each belief has been attained respecting the validity of the phenomena purporting to be 'caused by spiritual beings, together with the belief thence arising of the intercommunion of the living and the so-called dead.'"[34] Psychical researchers may have been more cautious in deciding whether extraordinary phenomena were genuine, but because they would potentially validate an occurrence as being the product of real spirit or psychic forces, they were unscientific and forwarded false information. Clodd continued,

"Analysed under the dry light of anthropology, [the society's] psychism is seen to be only the 'other self' of barbaric spiritual philosophy 'writ large.' It disguises the old animism under such vague and high-sounding phrases as the 'subliminal consciousness,' the 'telepathic energy,' the 'immortality of the psychic principle,' the 'temporary materialisation of supposed spirits,' and so forth."[35] Would any credible scientist support such absurdities? Figures such as Wallace and the physicist Oliver Lodge (1851–1940) lost much of Clodd's respect by supporting such ridiculous research. He concluded by arguing that even if these phenomena were proven to be real—which he thought was highly unlikely—he struggled to see what value it would have to science and society more generally.[36]

Unsurprisingly, Lang was quick to respond to the criticisms forwarded by Clodd, which he published in the journal *Folklore* later that year. As a leading proponent of telepathy and an active member of the Folklore Society, Lang could not allow these critiques to go unchallenged. Thus, he began his response by scrutinizing the problems with Clodd's condemnation of Wallace's and Lodge's scientific credentials. According to Clodd, these men were fooled by supposed spirit and psychic forces because their genius was restricted only to their knowledge of natural history and physics respectively. Outside of these domains, their ability to critically dissect evidence was little better than any other typical séance sitter.[37] However, who then was to be trusted as a credible observer? Lang argued that "men of science" were the obvious people to lead investigations—especially if they had attended the number of séances both Wallace and Lodge had. However, if they struggled in this task, the only other option would be to recruit the services of professional stage magicians.[38] Lang wrote, "The obvious plan is to try professional experts in conjuring, like Mr. Maskelyne, and I have reason to hope that this gentleman (whose logical fairness of mind rivals his ingenuity) may yield his assistance. Suppose, for argument's sake, that *he* is also puzzled; to what possible kind of observer would Mr. Clodd have recourse? Would he say 'To nobody the whole affair is nonsense'? If so, I fail to agree with him."[39] Even if a magician, such as John Nevil Maskelyne (1839–1917), failed to fully refute the spirit hypothesis and telepathy, it would still not be a closed matter for Lang, and he argued that "a performance—or, if you please, a series of occurrences—which neither a crowd of men of science nor an expert in conjuring could account for, would, I think, be a phenomenon worthy of examination."[40]

Lang continued his defense of spirit investigations by responding to Clodd's assertion that psychical research was nothing more than "old animism" in disguise. In Lang's view, that was a mischaracterization of the research field because psychic investigations were conducted with the aim not of supporting a belief in the existence of spirits per se, but of determining whether the evidence in support of genuine spirits was credible. More often than not, other causes could be identified as the real source of unusual circumstances, and surely rationalizing misbelief was something that anthropologists and folklorists should do. Take, for instance, Clodd's "sneers at the practice of looking into glass balls."[41] Both Lang and Clodd agreed that what was likely causing people to see supposed spirit phenomena in crystal balls was nothing more than hallucinations, created by some sort of hypnosis. If that was the case, then its explanation under animistic theory seemed highly justified. Lang asked, "Granting, then that such hallucinations exist, why on earth should they not be studied like any other mental phenomenon?"[42] There was ample evidence to show that seeing visions in crystal balls was both an historical and a global occurrence. Lang wrote, "The anthropologist, the folklorist, meets this practice and belief of 'scrying' in all ages, and among races of every grade of culture."[43] Researchers should therefore pursue this matter, because crystal gazing—or scrying, as it is also known—was an interesting cross-cultural survival from a bygone age.

What about table turning? Lang argued that this supposed manifestation had also been "burdened by the explanations of 'the old animism.'" But assuming that Michael Faraday and William Benjamin Carpenter were correct in their arguments that the phenomenon was caused by unconscious muscular motion, then once again this should be a topic pursued by scientific researchers.[44] Why are humans around the world and in all ages susceptible to automuscular movement? This repeated occurrence generated interesting questions, and Lang argued that "the phenomenon, in that case, of unconscious muscular action deserves study."[45] The same logic could be applied to other so-called spirit and psychic phenomena as well: if it is recorded in all ages and in all cultures, then it is of intellectual interest to researchers, who can use this material to further broaden understandings of human belief, history, and perception.

In his closing remarks, Lang refuted Clodd's assertion that anyone who studied spirit phenomena was likely nothing more than a credulous, superstitious fool. For example, Tylor was compelled enough by the testimonies

of witnesses professing to have observed genuine spirit and psychic manifestations to attend séances himself for the purpose of determining whether
there was any merit to the claims made by spiritualists. His trip to London
in 1872 was testament to the fact that the matter was sufficiently important
that it deserved direct investigation. Lang wrote, "I call him [Tylor] not
'superstitious,' but 'scientific,' and if nothing occurred, that does not make
him more scientific or less superstitious."[46] Whether Tylor's investigation
produced conclusive material did not matter—what was important was that
he recognized the integral value of prima facie evidence for maintaining
his scientific position. If he were to prove that spiritualism was a survival
of primitivism, the most compelling evidence was firsthand observation.

Predictably, Clodd published a derisive reply to Lang's defense of psychical research, and opened by stating that "Mr. Lang's 'Protest' is, in the
main, directed against my statement that modern psychism is but savage
animism 'writ large.' It is that statement which I have to attempt to justify."[47]
Clodd argued that his main objection to psychical researchers was not that
they supported the validity of the spirit hypothesis, and, he wrote, "they
can believe what they like."[48] His real problem was with how they collected
their data and critically analyzed their findings. In other words, it was not
a question of belief, but one of practice. The methods used by psychical
researchers, in Clodd's view, were highly flawed and unscientific. He argued, "My objection is *not* to the *research*, but to the *method* of it, which
under the guise of the scientific, is pseudo-scientific."[49] If psychical research
was done properly and analyzed under a more critical lens, then the only
conclusion to draw would be that belief in spirits is a form of animism. Any
witness professing to have observed genuine instances of spirit and psychic
phenomena was little more than superstitious nonsense or misperception.

A key example was the results produced by the SPR's committee on
hallucinations from 1894. At the founding of the society in 1882, a group of
leading members, including Henry and Eleanor Sidgwick, Frederic W. H.
Myers, and Frank Podmore, organized a team of investigators to interview
thousands of people from across Europe, to determine whether or not they
had either seen or felt something unusual that "appeared not to be 'due to
any external physical cause.'"[50] Of the 17,000 people interviewed by the SPR's
investigators, only 1,684 respondents (roughly 10 percent) could not account
for the causes of these phenomena. Even more striking was that 25 percent
of this marginal group were retelling the reports of other observers, thus

eliminating themselves as credible sources. With this aberration deducted from the total, only 1,263 respondents actually claimed to have directly witnessed inexplicable occurrences. Yet despite collecting only a minimal amount of affirmative responses, the SPR's committee on hallucinations argued that there was sufficient evidence to support "'the argument for the continuity of psychical life, and the possibility of communications from the dead,' while it unanimously holds it proved that 'between deaths and apparitions of the dying persons a connection exists which is not due to chance alone.'"[51] Clodd believed that this was an illogical conclusion, and demonstrated that the SPR were incapable of properly analyzing their data. He argued, "Does anyone in his senses believe that had this evidence been sifted *au fond* by a Committee of the Royal Society, or of the Folklore Society, such a conclusion as the foregoing would have been reached?"[52]

Who were these observers, anyway? Could one even trust them as credible witnesses of supposed spirit and psychic phenomena? Clodd argued that many of the people who were interviewed by SPR investigators possessed questionable moral characters, and low intelligence. Therefore, their responses should not be included in a scientific study. For a witness to be deemed a dependable and competent observer of alleged spirit phenomena, the investigator needed to consider carefully their suitability as a scientific subject. That raised another problem for Clodd, as he believed that many of the investigators collecting the testimonies for the SPR were unqualified. There were four hundred investigators supplying information to the committee, and some of them were devout spiritualists.

Take, for instance, the British journalist and editor William T. Stead (1849–1912), who supplied nearly one-tenth of the responses.[53] His status as a reputable writer rose during his editorship of the *Pall Mall Gazette* between 1880 and 1889. He then moved on to become the editor of the *Review of Reviews* in 1893. However, in that same year he also founded the spiritualist periodical *Borderland*. Its production was a way for Stead to combine his two interests of psychical research and journalism. Stead was utterly convinced that spirit and psychic forces were real, and even claimed to possess mediumistic powers. His assistant editor was the well-known clairvoyant Ada Goodrich Freer (1857–1931). In 1894 the SPR commissioned Freer to collect testimonies on second sight in the Scottish Hebrides. In Clodd's view, this was an example of how the SPR's research activities were conducted in a biased manner. How could investigators such as Stead and

Freer carry out objective interviews if they were so invested in the spiritualist movement? Clodd believed that their data as completely inadmissible.[54] He also argued that many of the other investigators lacked the requisite skillset and knowledge to produce data of a sufficient caliber for science. He wrote, for example, that "one-tenth of the collectors was drawn from classes not highly educated, [such] as small shopkeepers and coastguardsmen."[55]

If the quality of the investigators was not bad enough, Clodd also questioned the suitability of the committee itself as interpreters of the data. None of them in his view were sufficiently trained in science. Henry Sidgwick was a philosopher, and Myers was a classicist. Podmore was primarily a psychic investigator, but even so, he was not thoroughly enough acquainted with either the natural or physical sciences to be qualified to make a sound scientific judgement. Clodd disqualified Eleanor Sidgwick and her personal secretary Alice Johnson simply because of their gender. This is particularly striking, because of all the committee members, the two women were the most scientifically accomplished. Eleanor Sidgwick had collaborated with John William Strutt, Lord Rayleigh (1842–1919) on some physics research, and Johnson had trained as a zoologist. She even worked as a demonstrator in animal morphology at the Balfour Biological Laboratory for Women at the University of Cambridge between 1884 and 1890.[56] There is no doubt that Clodd's dismissal of their qualifications was strategic, but also bigoted. He was simply looking for a reason to undermine their credibility as dependable investigators of spiritualism. Fundamentally, because all of the committee were open-minded about the possibility of spirits and psychics being real, this eliminated them as reliable researchers. Clodd would have preferred "five thoroughgoing skeptics to Professor Sidgwick and his wife, Miss Alice Johnson and Messrs. Myers and Podmore."[57] In sum, from top to bottom, Clodd believed the whole process was poorly organized and executed.

The idea of even considering the possibility that spirits and psychics were real was laughable to Clodd. The findings produced by the SPR throughout the thirteen years of its existence was testament to "the fatuousness and imbecility—of the contributions which psychical research . . . [had] made to our knowledge of a spirit-world." Anthropologists had already discovered why all humans from every culture and age claimed to have observed genuine spirit and psychic phenomena. It was all the result of animistic belief, and Clodd stated, "Surely the identical character, *mutatis mutandis*, of apparitions, hallucinations, and the like among savage and civilized people, should make them [psychical researchers] pause to ask if the 'old animism' is not the chief factor in their production and persistence."[58] Any further investigations would only strengthen the case for animism as the most likely explanation for spirit and psychic phenomena. It was an open-and-shut case, and all psychical researchers were doing was fueling the fire of superstition, by leaving open the prospect that spirits and psychics might one day be proven to be genuine—an implausible feat.

Clodd's attack on modern spiritualism and psychical research during the late 1890s was tied to his staunch adherence to animism as the most likely explanation for the reasons why humans believed in the existence of spirits and psychics. However, any rational investigator could easily dispose of a testimony that claimed to have witnessed real spirit or psychic phenomena. A careful reading of the account would ultimately show that the observer was misinterpreting the occurrences either because of some illness (physical or cognitive) or as a result of trickery or hallucination. Clodd maintained this stance for the next couple of decades; the publication of *The Question* in 1917 was a continuation and expansion of this argument. Using the reports of witnesses dating back to the earliest years of the modern spiritualist movement in the late 1840s, Clodd wanted to prove once and for all that the belief in spirits and psychic forces was a survival and revival of primitive superstition.

Questioning the Spirit Hypothesis

A close reading of Clodd's book reveals the processes by which he critically evaluated his sources and constructed his visual epistemology.[59] It was a staunch critique of spiritualism and psychical research, written by a hardline disbeliever, and it should therefore be read with this bias in mind. Who could be trusted with making reliable observations was of key concern to Clodd. In the introduction to *The Question* he outlined some of the problems all scientists faced when observing different kinds of phenomena. Take as an example astronomy, a science that was highly reliant on "skilled vision."[60] He wrote, "In astronomical observations absolute accuracy is impossible, because eyes and other conditions vary in each observer: hence variation in the reports which each brings. To arrive at a sure result, there are made such additions to, or subtractions from, a number of observations of the same celestial object as will compensate for known causes of error."[61] This tendency toward misobservation is known as the "personal equation," a term generally denoting the allowances given for individual bias and idiosyncrasy.[62] Because of the extraordinary nature of spirit investigations, which relied heavily on personal testimonies that were riddled with errors and misinformation, researchers studying the subject had to be particularly vigilant in identifying possible anomalies. To test the veracity of an account, investigators were best placed to use a form of "collective empiricism" to determine whether the unusual phenomenon witnessed by an observer was comparable to other similar circumstances.[63]

The mind was also prone to playing tricks on observers, especially when witnessing phenomena that were as abnormal as alleged spirit and psychic forces. According to Clodd, "normal" minds have their "fallacies," while "abnormal" ones have "delusions and illusions," which misdirect human perception. However, to complicate matters more, the mind also has "stored-up myriads of impressions which have passed unheeded by us into our potential consciousness, and which become active under various, often abnormal, mental states," further influencing perceptual awareness.[64] With all these cognitive challenges, it is no wonder that so many people believed that they had seen genuine spirits and psychics. Other sciences such as physics and chemistry had experimental knowledge to back up incredible observational claims, and, as Clodd remarked, "when after repeated tests, the results anticipated by the theory are found to be unvarying, the theory is established."[65] However, because of the unpredictable nature of spirits and psychic forces, reliable tests were few and far between in spirit investigations. Thus, as Clodd argued, "We have to accept or reject what Spiritualists tell us, and supplement this, so far as we can, by observations made . . . under difficulties not attending to [other] branches of research."[66]

The observations collected during spirit investigations also suffered from animistic interpretations that were a survival of primitive ages. All humans were influenced by these core superstitious beliefs, and this was an argument that Clodd had been articulating since the closing decades of the nineteenth century. He stated, "At the outset of the inquiry, a hearing must be accorded to what the anthropologist has to say on the pedigree of Spiritualism. We shall learn from him that this pedigree stretches into a dim and dateless past, reaching to the animistic stage in the evolution of religion: a stage when men conceived of spirits indwelling in everything." This core belief in supposed spirits continued to exist even into modern society, and although so-called civilized people should know better, they were incapable of fully escaping animistic belief. Quoting the classicist Gilbert Murray (1866–1957), Clodd argued that "the mind of man cannot be enlightened permanently by merely teaching him to reject some particular set of superstitions."[67] So long as other cultural fantasies continued to circulate in a society, the mind would always be drawn back toward the irrational. The only defense against devolving back to superstition was to show continually that natural and mechanical causes produced extraordinary phenomena. Observers who were fooled into thinking that an occurrence was the result of genuine spirit or psychic forces lacked critical awareness, and were regressing to a "primitive state."

Environment had a huge impact on misperception and misbelief. The origins of modern spiritualism emerged within a highly religious context that was prone to superstition. The movement began in western New York, a region that was historically populated by various Christian evangelical groups. It had been described as the "burned-over district," and the place of the "second great awakening," because of the force by which religious revivalism had passed through it.[68] In an environment swelling with spiritual fervor, local inhabitants were predisposed to accept the early claims of mediums and their followers as true. The first well-known case of modern spiritualism is usually attributed to the Fox sisters Leah, Margaret, and Kate, who lived in western New York during the peak years of Christian revivalism. Had the girls resided in another location, less engulfed in religious intensity, the fate of the movement might have been different. However, according to Clodd, it was clear that modern spiritualism was the product of "old animism," under the guise of Christian revivalism.[69] Social contexts such as the burned-over district were responsible for producing reports that supported the spirit hypothesis. However, any avowed case of genuine spiritualism, observed under the critical eye of naturalism and rationalism, would eventually be exposed as nothing more than a misinterpretation of the circumstances. Clodd devoted the remainder of his book to assessing the legitimacy of professed spirit and psychic encounters. If he could not fully refute the claims made in an account, he would create enough doubt to undermine the trustworthiness of the source's value for legitimizing spiritualism.

Doubting the Sources and Exposing the Fraud

One of the high-profile cases that Clodd examined in *The Question* was the Cock Lane ghost story. A key reason he wished to revisit the case was to repudiate the revised version Lang had developed in *Cock Lane and Common Sense* in 1894, where he suggested that the whole affair was a case of unrecognized telepathy. Clodd rejected the argument that the results of the inquiry were debatable because of the alleged coercion used by the investigators when examining Elizabeth Parsons. In his view, there was more than sufficient evidence to show that the whole situation was nonsense. Clodd's evidence came from the same sources Lang used: the pamphlet written by Oliver Goldsmith; reports in periodicals, such as the *Gentleman's Magazine*; and miscellaneous tidbits found in the British Museum's catalogue.[70] The main difference between Clodd's and Lang's respective versions of the story was that Clodd's account included far more examples of observations made by skeptics and disbelievers.

One of those skeptics was the antiquarian and historian Horace Wallace, Earl of Orford (1717–1797), who wrote a letter to his friend the diplomat, Horatio Mann (1706–1786), on January 29, 1762, about his impending visit to the apartment on Cock Lane. Before even seeing the phenomena firsthand, the Earl of Orford was already skeptical about their legitimacy. He stated, "We are again dipped into an egregious scene of folly. The reigning fashion is a ghost."[71] His doubts were confirmed, and four days later in another letter he described his encounter. Richard Parsons was depicted as a drunken parish clerk "out for revenge," and the apartment was hardly an ideal place for conducting a rigorous investigation of psychic or spirit phenomena. As a result of the popularity of the story, the Parsons family were flooded with visits. Therefore, the room was constantly overcrowded with as many as fifty people at a time. It was also poorly lit, making it easy for trickery to occur. However, what really caught the attention of Lord Orford was the monetary gains that both the Parsons family and the local businesses were earning as a result of the affair. He wrote, "The Methodists have promised them [the Parsons] contributions; provisions are sent in like forage, and all the taverns and ale-houses in the neighborhood make fortunes."[72] The whole situation seemed to be a profitable ruse, and Lord Orford concluded his letter by stating that "this pantomime cannot last much longer."[73] It was a damning testimony, and one that undermined the version outlined in Lang's book.

The real problem for Clodd, however, was whether the investigation led by Samuel Johnson was credible, or if he used excessive pressure to force an exposure. Clodd believed that Johnson acted judiciously throughout the process, and the various controls that he put in place were highly credible.

Prior to even discovering the little wooden board that was concealed in Elizabeth Parsons's nightdress, Johnson reported that the investigators suspected that she had "some art of making or counterfeiting a particular noise, and that there is no agency of any higher cause."[74] It was merely a matter of discovering how she did it. Moving Parsons to a "neutral site" that she had never visited, nor held any ties to, was good practice, and allowed for an objective investigation. Having multiple witnesses on hand to corroborate testimonies, making sure the space was adequately lit, and requesting that the girl's hands remained visible at all times were also sensible precautions against trickery. It limited the possibility of the girl manipulating the space during the observations.[75]

When more days passed without witnessing any spirit phenomena, the investigators grew frustrated, and threatened to commit Parsons to Newgate for imposture, if she was unable to produce some positive results by the following night. It was only then that noises occurred, and she was "searched and caught in the lie."[76] Supporters were quick to argue that the threats were the real reason why Parsons cheated, and that if she had been given more time, genuine spirit noises would have been produced. Clodd, however, argued that this was just a tactic used to preserve her reputation. She was caught because she did not actually possess mediumistic powers, and extending the investigation for more days would not have yielded different results. The Cock Lane ghost was simple trickery. When Lang restated that the investigators used bullying tactics and coercion in his book, he fell victim to an old false rhetoric. There were insufficient grounds for the case's legitimacy, especially given the overwhelming evidence in favor of Parsons faking the manifestations. Where was the conclusive evidence favoring telepathy as the probable cause? There was none, according to Clodd, and therefore Johnson's conclusions were correct. The case of the Cock Lane ghost had been weighed carefully, and Clodd believed that the investigators correctly appraised the case in the 1760s. It was not an example of genuine spirit or psychic forces, and Lang's revisionist account was erroneous.

Clodd also made a strong case for Johnson as a credible investigator of spirit and psychic phenomena. He had a talent for critically evaluating extraordinary cases, and identifying the likely causes of unusual circumstances. He was, therefore, not a man who was easily fooled by misperception or superstition. For instance, Clodd recounted a story in which Johnson explained the causes that produced an alleged haunting at the Hummums

Hotel in Covent Garden, London. A waiter working in the hotel claimed to have encountered a guest while walking down into the cellar. When he told his boss about the matter, he was informed that this particular guest had recently passed away. Terrified by this revelation, the waiter soon fell ill with a high fever. However, Johnson argued that the waiter was most likely already suffering from illness when his supposed encounter with the deceased guest occurred. The event was nothing more than a hallucination brought on by the fever. Clodd argued, "Given a healthy condition and body, there is no room for phantasms of either the living or the dead. The causes which beget them are explained and their doom is certain."[77] Johnson's composure in carefully analyzing the story and providing a reasonable account for why the waiter saw an alleged spirit demonstrated his skill as a dependable and trustworthy investigator of extraordinary phenomena. His poised mind allowed him to identify the most likely cause of the unusual circumstance. Given his abilities, Clodd argued that Johnson's conclusions regarding the Cock Lane ghost should be trusted as well.

Of all the celebrity mediums of the nineteenth century, Daniel Dunglas Home ranked among the best. What made him stand out above all the others was that he could allegedly produce an astounding range of incredible feats under a range of strict controls. Most reports claimed that his performances were done in good lighting, and, more significantly, he was never exposed as a fraud. His reputation as a trustworthy medium, willing to be investigated by anyone who wished, did much to strengthen his legitimacy.[78] As Podmore argued, "the main defenses of Spiritualism must stand or fall" with him.[79] However, Clodd was suspicious of Home. He may not have been caught cheating, but something about his character seemed dishonest. Home's ability to acquire trust in his supporters was linked to the type of personality he self-fashioned. He claimed to have never taken money for his performances, but he received many gifts from his patrons. Much like Tylor in his assessment of Home from his notebook in 1872, Clodd believed that the medium was a clever con man.[80] Considerable space is devoted in *The Question* to raising suspicion of Home's moral integrity.

Home's rise to fame among spiritualists came only a couple of years after the events of the Fox sisters in 1848. By the mid-1850s, he had already amassed a large group of followers, who even paid for his journey to England in 1855. From there, he traveled across the European continent, performing for all sorts of wealthy audiences. Eventually, he married Alexandria de

Kroll (1841–1862), a young Russian noblewoman. With his newfound wealth, he returned to England in 1859. Tragedy struck, and his wife died in 1862, leaving him with little economic stability. He relied heavily on the kindness of wealthy patrons, who often allowed him to live for free in their homes. Finally, in 1866, after much hardship, he met a wealthy widow named Jane Lyons, who became his chief patron. Clodd sarcastically remarked that Home "won her heart and opened her purse strings."[81] So amazed was she by his psychic powers that Lyons bequeathed Home thousands of pounds. Although some reports claimed that her gift was made voluntarily, Clodd suggested that Home might have used some sort of "undue influence upon her," such as hypnosis, to gain access to her wealth.[82]

Soon after giving Home the financial capital, Lyons "cooled and repented, and brought an action for the restitution of the money, which she won." The court acquitted Home of any wrongdoing, but the whole affair raised questions regarding his moral character. Why did Lyons have such a dramatic change of heart? What could have caused her to end the friendship with Home? For Clodd, this indicated that something was not quite right about the medium. His well-crafted reputation was obviously masking some nefarious activities. Why else would his benefactor rescind her economic support? If a close friend such as Lyons distrusted Home, why should someone who barely knew him trust him at all? Home continued to depend on wealthy patrons for his livelihood until 1871, when he married his second wife, Julie de Gloumeline, another wealthy Russian noblewoman, who supported him until his death in 1886.[83] Home's reliance on wealthy women to support his lifestyle made Clodd suspicious, and this way of life did not seem to Clodd like one belonging to the same man whom many spiritualists framed as a person of high moral standing.

If it were so easy to raise doubts over Home's moral integrity, then surely one could also find inconsistencies or potential errors in the testimonies of witnesses who had claimed to see him produce extraordinary phenomena. Clodd believed that Home was almost certainly being deceitful at his performances as well, and although he had never been exposed as a fraud, it was significant that he "always chose his own company or imposed his own conditions." When arranging the seating at a séance table, "he assigned each one his place; [and] those who had the greater faith in him were rewarded by being put nearest to him."[84] That was highly suspicious, according to Clodd, because it afforded ample opportunity for Home to cheat during séances.

For a fair inquiry, investigators needed full control over the space where the phenomena took place, and the best seats to observe his performances. As for the alleged good lighting at his séances, Clodd remarked that "the stock phenomena of raps, tilting tables, . . . spirit voices and spirit lights" occurred in dim lighting, but when the conditions were "judged favorably to the higher manifestations, the lights would be turned out."[85] Thus, solely in total darkness would spirit materializations and levitations occur. This was a very different representation of Home's performances than the spiritualist literature usually portrayed.

Clodd also questioned the veracity of the testimonies supporting the genuineness of Home's psychic powers in other ways. Take, for example, Home's famous levitation in 1868 at Ashley House in Victoria, London.[86] One of the key witnesses to the event was Alexander William Crawford Lindsay. Clodd argued that such a man could not be trusted as a credible witness because "he was subject to hallucinations of black dogs, figures of women, and flames of fire on his knees." A mind susceptible to visions could not be counted on for confirming the reality of psychic phenomena, and caution was required when accepting Lindsay's "testimony to the suspension of the law of gravitation."[87] Moreover, if Lindsay likely hallucinated Home's levitation, then the other witnesses probably did, too. Given that Home had control over the space, he could have also set up an illusion in advance of the performance. After all, no descriptions of controls were ever given in the accounts of the event at Ashley House.

Clodd was equally critical of William Crookes's investigation of Home during the summer of 1871. Crookes had performed experiments to determine whether or not Home had the ability to alter his height and weight. After rigorous testing, he concluded that the medium possessed some "hitherto unknown force" that was likely responsible for the feat.[88] Clodd, however, contended that there were inadequate controls in place to ensure an objective investigation. Most notably, "Home again prescribed the conditions of the experiment," affording him ample opportunities to influence the findings.[89] Clodd also argued that Home was highly dexterous, and given Crookes's poor eyesight, could easily have manipulated the measuring apparatus without being caught.[90] Other witnesses also attested to Home's elongation abilities, but Clodd contended that the "extent of this is reported to have varied at different times from four to eleven inches."[91] With so many inconsistencies in the measurements, the evidence was far too suspect to use as confirmation of genuine mediumistic powers. Lord Lindsay also attested to seeing this elongation feat, but as Clodd argued, "the reader must judge himself" whether his lordship's observations could be trusted, given his penchant for seeing visions.[92]

Another medium to be re-evaluated critically by Clodd was William Stainton Moses. As a respected university educator and former Anglican minister, Moses garnered much trust from his séance sitters. He never charged a fee for his mediumship, and he often took a financial loss when traveling to different homes. Even more praiseworthy was that Moses had severe respiratory problems caused by tuberculosis, and yet he continued to perform under the poorest of health. As Clodd remarked, "Moses inspired implicit confidence in his integrity."[93] Many leading psychic investigators were also impressed by Moses's mediumistic abilities. Wallace, for instance, stated that he was "as remarkable a medium as D.D. Home," while Lang claimed that Moses produced more types of extraordinary phenomena than any other known psychic throughout history.[94] However, these admirable qualities and glowing endorsements were not enough to convince Clodd that his powers were real. The evidence in support of his mediumship had too many problems. Much like Home, more uncertainties surrounded Moses's career as a psychic than the reports typically suggested.

Take, for instance, Moses's mental constitution. Clodd argued that "Moses was a neurotic, [and] therefore of highly susceptible temperament."[95] He may very well have believed in the genuineness of his powers, but he was likely suffering from some sort of psychosis. He was also an alcoholic, which Clodd believed further demonstrated that his "inhibitory power" was weak.[96] Those attending his séances would have picked up on this unsettled energy, and it likely fostered a sympathetic environment, conducive to accepting the reality of his mediumship. Although there was "accurate and systematic records of all the phenomena" that occurred during his séances, the accounts were highly dubious.[97] Moses was extremely selective about who could sit at his séance table, and typically chose a small circle of intimate friends, who were all loyal spiritualists. Thus, their testimonies hardly represented "objective" reporting. Witnesses claimed that Moses was able to produce a broad range of phenomena at his séances, including "rapping alphabets" and "diffused fragrances." However, these occurrences happened only under total darkness, because Moses "wholly excluded" light from his séance rooms. The rationale was that spirits preferred to operate in darkened spaces, but Clodd found this argument to be a highly convenient justification for allowing Moses to create an environment favorable for deception. Given the number of problems with the evidence supporting the veracity of Moses's mediumship, Clodd wondered how anyone could believe in its authenticity.

In one of the more famous cases of Moses's mediumship, it was alleged that he suddenly fell into a psychic trance, and sensed the death of a stranger, who "threw himself under a steam-roller in Baker Street." The main witness of his spiritual convulsion, which led to the revelation, was an unnamed spiritualist, who reported that in an agitated state Moses "drew a rough sketch of some horsed vehicle and then wrote: 'I am killing myself today, Baker Street.'"[98] William Fletcher Barrett was utterly convinced of the reality of this affair, and believed that the deceased man's spirit had visited Moses to deliver the news. Clodd, however, was of another opinion. Because the main witness of Moses's trance was another unnamed spiritualist, that in itself was enough to disregard the story altogether. Moreover, there were other important factors to consider as well. The suicide occurred in the neighborhood where Moses lived, and it happened early enough in the day that it was possible for him to know about the death even before the news was broadly circulated.[99] It was also irregular that the suicide victim's name was unknown to Moses. Surely if the spirit was actively seeking to announce its death to the medium it would identify itself by name. Remaining anonymous seemed misguided. Why even announce your death if the medium could not inform friends and family?

Clodd argued that this case seemed dubious, especially because thirty-eight other spirits communicated with Moses and "this one alone did not reveal his name."[100] Clodd had an explanation for this apparent anomaly, though. Most of the spirits to visit Moses were of deceased celebrities, about whom information was readily available. After Moses's death in 1892, it was noted that his library contained many biographies, and it was likely that Moses used this material to harvest personal information about the supposed spirits who visited him. It would have been significantly harder to provide personal details about an anonymous man committing suicide in Baker Street, however. It was for this reason that Clodd believed the victim's name was unknown to the medium. In the absence of any real psychic powers, the only information that Moses could have gathered in the immediate aftermath of the death was basic details from the personal testimonies of locals, who had witnessed the event.[101]

Figure 4.1. Image of the Davenport Brothers sitting in their spirit cabinet. Various instruments can be seen inside the center compartment, as well as the rope used to bind the brothers. At the top of the cabinet's center door, there is a diamond-shaped hole where spirits' hands and feet allegedly appeared during performances. *Source*: Henry Ridgely Evans, *The Spirit World Unmasked: Illustrated Investigations into the Phenomena of Spiritualism and Theosophy* (Chicago: Laird & Lee, 1897), 139.

There were always going to be cases of mediumship that remained unsettled, because the evidence was too patchy to form a decisive conclusion. However, if one could show that the overwhelming number of investigations examining a particular medium resulted in inconclusive findings, then the legitimacy of their powers would be greatly weakened. The only rational positions to hold would be that of skepticism or disbelief. Ira Erastus Davenport (1839–1911) and William Henry Davenport (1841–1877) were used by Clodd as an example for this very purpose—to illustrate doubtful mediumship. The brothers had risen to fame during the late 1850s as a result of their spirit cabinet performance (see fig. 4.1). Sitting inside a large wooden cabinet with three compartments, the brothers had their hands and feet tightly bound in front of an audience. Various instruments including bells, tambourines, a trumpet, and a guitar were placed in the center section of the cabinet, before the three doors were closed and locked. After a few moments, some of the instruments would begin playing, and at a small opening in the top center of the cabinet's middle door, spirit hands and feet would appear. Once the phenomena ceased, and the doors would be reopened, the brothers would be found as they had been left, with their hands and feet bound on opposite sides of the cabinet.[102] It was a hugely popular performance, but like so many spiritualist acts, its success eventually came to an end after the mediums were unable to produce phenomena under more rigorous controls.

For believers, the Davenports' performance was strong evidence in support of genuine spirit and psychic forces. Spiritualists believed that by entering into a trance state, the brothers were able to channel spirits, who played the instruments, and produced the phantom hands and feet. However, skeptics and disbelievers like Clodd found this explanation absurd. The spirit cabinet act was nothing more than an elaborate escapist trick. By twisting and turning their hands and feet, the brothers were able to free themselves from the bindings in order to play the instruments, and wiggle their limbs through the small opening in the cabinet's middle door. The whole act hinged on their ability to slip back into the bindings before the cabinet doors were unlocked to reveal them again. If it was all caused through spiritual agency, why close the doors in the first place? With a clear view of the brothers, the bindings would not even be necessary.[103]

To add further credibility to their spirit cabinet act, the brothers traveled with an American congregational minister named Jesse Babcock Ferguson (1819–1870), who vouched for their good character before each performance commenced. Clodd argued that these endorsements did much to strengthen their reputation. It was widely acknowledged that Ferguson was utterly convinced of the brothers' powers, and did not suspect any deceit on their part. The sincerity of his opening lectures fostered an atmosphere of open-mindedness that greatly benefited the Davenports' performance, lessening the level of suspicion. During their tours they were also accompanied by their manager and secretary, D. Palmer, and an assistant, William M. Fay (1840–1921), who was a skilled magician, and also allegedly possessed some mediumistic abilities. This was a useful accomplice to have on hand during a performance that professed to demonstrate genuine spirit phenomena.[104]

With the outbreak of the American Civil War, the brothers and their troupe sailed to Britain, where they were joined by the American journalist and devout spiritualist, Thomas Low Nichols (1815–1901). He helped promote the Davenports' shows to British audiences, and kept a detailed record of their extraordinary feats. He even published a biography of the brothers in 1864. Nichols was a fervent supporter of the Davenports, and argued in his book that they were missionaries for the movement. He believed that the British were far more skeptical and materialistic than Americans, and were "firmly settled in the belief that there is, and can be nothing beyond the range of ordinary experience."[105] It was, therefore, up to talented mediums such as the Davenports to open British minds to new possibilities. He wrote: "If the manifestations given by the aid of the Brothers Davenport shall prove to the intellectual and scientific classes in England that there are forces—and intelligent forces, or powerful intelligences—beyond the range of their philosophies, and that what they consider physical impossibilities are readily accomplished by invisible and to them unknown intelligences, a new universe will be opened to human thought and investigation."[106] In Nichols's view, the veracity of the spirit hypothesis stood or fell with the Davenports, and it is no wonder that Clodd was keen to emphasize how improbable their psychic powers were. Nevertheless, as with other cases, Clodd explicated how the evidence surrounding the Davenports' performances disavowed the genuineness of their mediumship.

In 1864 the Davenport brothers visited Liverpool, where they performed in front of a large crowd. Usually the brothers would have someone from their entourage tie the ropes around their hands and feet. On this occasion, however, two volunteers from the audience insisted that they be allowed to bind the performers; their names were Mr. Hulley and Mr. R. B. Cummins. They used a type of knot known as a "Tom fool's knot," which was particularly secure, and the brothers instantly recognized that they would be unable to escape their bindings. They pleaded with the crowd that the knots were too tight, and that they were losing circulation in their hands and feet. As Clodd remarked, "A doctor, summoned to give his opinion, said that the knot was not harmful." However, the Davenport brothers refused to perform. Enraged by their refusal, the crowd began accusing the troupe of being frauds.[107] The Davenport brothers left town soon after, but were followed by Hulley and Cummins. At every city that they traveled to, the two men would push to the front of the crowd and insist that they be allowed to bind the brothers. On one occasion they even incited a riot, leading to the crowd destroying the Davenports' cabinet. Forced to cancel their shows, and rebuild their cabinet, the brothers and their troupe fled to the European continent in search of less interrogative audiences.[108] These circumstances were highly suspicious to Clodd, and suggested that the Davenport brothers were almost certainly frauds.

While touring throughout Europe, the Davenports were joined by a devout spiritualist named Robert Cooper, who spent seven months traveling with them, and recording his experiences at their performances. Amazed by their extraordinary displays, he sought to re-establish their credibility in Britain. He began by attesting to their greatness in a biography titled *Spiritual Experiences*, which was published in 1867.[109] When Cooper returned to Britain that same year, he initiated a campaign for the restitution of the Davenports' reputation as genuine mediums. Cooper believed that if he could get a scientific body to investigate and confirm the brothers' powers as legitimate, such a study would acquit them of any previous mishaps. After multiple rejections from various scientific groups, he finally found an eager audience at the Anthropological Society of London (ASL). One of the figures to help him convince anthropologists to take on the case was Wallace, who saw the Davenports' potential acquittal as an opportunity to align psychical research with anthropology. Cooper had less lofty scientific ambitions, and merely hoped the investigation would resurrect the Davenports' career in Britain.[110]

In 1868 the brothers were finally persuaded to return to Britain and meet with the ASL, but the investigation did not go as planned. Before a thorough examination even commenced, the two parties were unable to reach an agreement on what controls were allowable during the séance. As Podmore later recounted in his book *Modern Spiritualism* from 1902, "The committee . . . offered to proceed with the investigation, on the conditions that they should themselves supply the ropes and other materials, should be allowed to hold the hands of the mediums during the 'manifestations,' to apply coloring matter to the hands either of mediums or 'spirits,' and to open the door of the cabinet as soon as a spirit hand appeared."[111] The brothers refused, and despite some haggling between the secretary of the ASL, J. Frederick Collingwood, who led the committee, and Cooper, the investigation was eventually abandoned. The details of these exchanges were published as a series of letters in the spiritualist journal, *Human Nature.*[112] Clodd argued that if the Davenports truly possessed psychic powers, there was no reason for them to reject these precautions. How did these controls inhibit the actions of the spirits? It was a missed opportunity for the Davenports to demonstrate their legitimate mediumship. Had the ASL been permitted to test them, and confirm no trickery was at work, there would be proper scientific evidence supporting the spirit hypothesis.

The final blow to the Davenport's credibility occurred a few months later, when the magician John Nevil Maskelyne and his friend and prop maker, George Alfred Cooke (1825–1905), perfectly reproduced the spirit cabinet act before a large audience at the Crystal Palace in London. Spiritualists who had attended the performance were so amazed by how closely it replicated the Davenports' feats that some, like the American spiritualist writer, Benjamin Coleman, believed Maskelyne and Cooke to be powerful mediums as well.[113] For Clodd this was further damning evidence that the Davenports were clearly fakes. If they genuinely possessed mediumistic powers, a professional conjuror such as Maskelyne would not have been able to reproduce their psychical feats. The only logical conclusion was that the Davenports were also talented and skillful illusionists. In sum, although they were not openly caught cheating, their refusal to be investigated transparently, combined with Maskelyne copying their performance flawlessly, raised far too many doubts over their legitimacy as mediums. Clodd, therefore, believed they were total charlatans.[114]

One of the most infamous cases of false mediumship was that of the American Henry Slade (1835–1905). His main performance centered on a type of spirit communication in which seemingly blank slates would somehow receive written messages, which Slade maintained came from the spirit world. It was a hugely popular act, and Slade's reputation within the spiritualist community rose quickly as a result. What made his performance particularly impressive were the apparent controls he put in place to ensure that everything was done completely aboveboard. As Clodd described in his book, the spirit communications "were produced in full light. The company were free to bring their own slates, mark them for identification, fasten them up, lay them on the table, each one keeping his or her eyes steadfastly on the medium."[115] Yet, because Slade had total control over the space, there was still ample opportunity for him to cheat, and as Clodd explained, Slade was exposed as a fraud on multiple occasions.

The first exposure occurred in New York in 1872. After sitting through two unsuccessful séances, a self-proclaimed pickpocket and fake medium named John W. Truesdell alleged that Slade had confessed to him that his whole performance was an act of deception. He used various methods employed by professional magicians to stage the phenomena.[116] However, these allegations did not gain much mileage. Because of his own indiscretions, Truesdell was hardly seen as a reliable witness for psychic investigations. Eventually, Truesdell published Slade's so-called confession in 1883, but because it bore too much likeness to the author's own writing style, many believed that it was untrustworthy as an evidentiary source to refute the legitimacy of Slade's mediumistic powers.[117] Nevertheless, it was still suggestive, implying that more was going on during Slade's performances than witnesses were observing. What was needed to confirm the illegitimacy of Slade's psychic abilities was an investigation led by a credible witness with suitable training in critical observation.

Such an investigation occurred in 1876. While attending one of Slade's séances in London, the biologist E. Ray Lankester (1847–1929) and the physician Horatio Bryan Donkin (1845–1927), grabbed an allegedly blank slate from the medium's hand, which contained a prewritten message on it. Slade was immediately denounced as an impostor.[118] Lankester and Donkin published their findings in a series of scathing letters for the *Times* in the autumn of 1876, where they described the whole affair in detail.[119] Circumstances worsened quickly for Slade, and he was prosecuted for fraud, and sentenced to three months in prison. However, a minor oversight with the wording used in the original court summons, afforded Slade an opportunity to appeal, and while waiting for the new paperwork to be processed, he fled back to the United States to evade any jail time. Nonetheless, the sensation surrounding the whole event ruined his reputation in Britain. The damning testimonies of Lankester and Donkin were confirmation that Slade's performances were nothing more than clever tricks of sleight-of-hand and misdirection. Clodd was thoroughly convinced that the Slade affair was strong proof against spiritualism.[120]

A similar account was given of Florence Cook's fall from grace. Like Slade's, her rise to fame among spiritualists was quick but short-lived. Cook was widely recognized by spiritualists as one of the greatest spirit materializers in Victorian Britain, and her performance centered on the manifestation of the spirit Katie King. At a typical performance, Cook would either retire into an adjoining room or enter a large cabinet, where she would cover her head with some sort of shroud, before entering into a trance. The doors would then be shut, but inevitably reopen a few minutes later when King would emerge to interact with guests. Usually someone appearing to be Cook would continue to lie covered, but in view, while the performance went on. Afterward, the doors would close again, before Cook would eventually re-emerge to speak with the sitters. Although she received a glowing endorsement from William Crookes in May of 1874 attesting to the genuineness of her psychic powers, it was significant that she had already been caught cheating at a séance in December of 1873.[121]

It was reported that while the so-called spirit Katie King walked around the room, a suspicious guest recognized an uncanny resemblance between King's physical features and those of Cook. He grasped the spirit by the hand in the hope of exposing the whole performance as a sham. Much to his dismay, a couple of Cook's devoted followers came to the rescue, and, as Clodd recounted in his book, "Katie retreated to the cabinet, which, after a delay of five minutes, was opened, revealing Miss Cook, dressed in black and seated."[122] For Clodd, this was fairly damning evidence, showing that the act was not a genuine example of a spirit materialization. However, many spiritualists rubbished the event as a purposeful attempt to discredit Cook by a rival medium. The guest who had grabbed the alleged spirit was the lawyer William Volckman, a close friend and future husband to the medium Agnes Guppy. A rivalry had recently formed between the two mediums as a result of Cook's sudden growth in popularity. Fewer people were attending Guppy's performances, generating much jealousy and animosity. Cook's defenders contended that Volckman's actions broke with séance protocols, and therefore should not be considered as genuine evidence against the legitimacy of her mediumship.[123] Although Cook's reputation just about survived the ordeal, she suffered a further blow to her credibility in 1880, when the antiquarian and politician Sir George Sitwell (1860–1943) also showed that Cook and King were the same person during one of her séances. This was absolute confirmation to Clodd that Cook was a total fraud.[124]

In addition to weaving a rich history of imposture among modern mediums, Clodd also discussed other types of spiritualist paraphernalia that were exposed as fake throughout the nineteenth century. A notable example was spirit photography, a practice in which photographers used seemingly ordinary cameras and photographic plates to capture images with supposed spirit extras in them.[125] The first spirit photographer to gain widespread acclaim was the American William Howard Mumler (1832–1884). During the early 1860s he ran a highly successful business in Boston, selling so-called genuine spirit photographs to an enthusiastic clientele. Often these images were of well-known mediums, such as Charles H. Foster (1838–1888) (see fig. 4.2).

Originally from Salem, Massachusetts, Foster was known for a type of phenomenon called "skin writing," in which messages from the deceased would suddenly appear on part of his body. He also performed a feat known as pellet reading, in which sitters would write the names of deceased people on slips of paper, which would be rolled into pellets and put into a bowl. The medium would then pick one randomly, and without opening it, speak about the deceased person's life. Foster was exposed as a fraud in 1872 by Truesdell, who figured out that he used basic misdirection techniques to switch the pellets. At the time of his photographing, however, Foster still maintained a reputable status among spiritualists.[126] Nevertheless, for later observers such as Clodd, Foster's exposure as a fraud only served to further highlight how unreliable spirit photographs were as sources that supported the genuineness of spirit phenomena. Any kind of extraordinary occurrence associated with Foster was naturally greeted with skepticism. Given that Mumler produced the spirit image with Foster, it raised doubts about the photographer's credibility as well.

Mumler's good fortunes came to a swift end when another local spiritualist, a physician named H. F. Gardner, argued that many of the alleged spirits in the photographs were recognizable Bostonians, who were still alive. Gardner's declaration was accepted by most observers—he was a highly respected member of the spiritualist community, who regularly delivered public lectures on the topic. His training in medicine also strengthened his credibility as a trustworthy exponent of genuine spirit phenomena. Mumler was completely ruined by the claim, and in an attempt to rebuild his reputation he relocated to New York City.

▶ Figure 4.2. A photograph taken by William Howard Mumler of the medium Charles H. Foster during the 1860s. The outlines of an unknown spirit can be seen standing directly behind Foster, with its arms wrapped around his neck. *Source*: "Charles H. Foster," by William Howard Mumler, albumen silver print, photographed in Boston, MA, 1862–1875, catalogue number 84.XD.760.1.9, J. Paul Getty Museum, Los Angeles.

Things were not much better after the move. He made several powerful enemies in New York, including the famous American impresario P. T. Barnum (1810–1891). Mumler was eventually charged with fraud, and during the trial, Barnum acted as a star witness for the prosecution. Barnum argued that Mumler was a swindler. He was duping vulnerable people, suffering from grief, into paying high fees for photographs supposedly containing evidence of their deceased loved ones continued spiritual existence after death. A duplicitous business such as Mumler's photography studio had to be closed down immediately. However, as Clodd recalled in his book, Mumler "got off owing to a technical defect in the indictment," which was due to an insufficient amount of evidence showing his guilt.[127] He was acquitted of any wrongdoing, but his career as a spirit photographer was over. Nonetheless, if the alleged founder of spirit photography was a complete charlatan, what did it say about those who followed in his methods? Clodd believed it was a clear indication that the whole enterprise of photographing spirits was utter nonsense.[128]

The Future of Animistic Theory

As a staunch adversary and disbeliever of the spiritualism and telepathy, Clodd had carefully assessed dozens of famous cases of mediumship to determine the veracity of the spirit hypothesis. The evidence that he presented was overwhelmingly against the argument that spirits and psychics were real. In most cases, the mediums or spirits were shown to be the result of simple trickery, or hallucination brought on by illness or intoxication. Even if the case was not outright disproven, Clodd argued that he provided sufficient grounds for dismissing it as firm support in favor of the legitimacy of modern spiritualism. The strength of animism had been tested, and it seemed to account for the causes of most cases of so-called spirits and psychic phenomena. Thus, for Clodd, the spirit hypothesis was nothing more than a survival and revival of primitive, superstitious thought.

Clodd continued to debunk spiritualism during the final decade of his life. His last major contribution to the dialogue were his two lectures on occultism, which he delivered at the RI in 1921 and published the following year.[129] However, by the start of the 1920s there were significant transformations occurring within British anthropology, and new disciplinary preoccupations started overshadowing animistic studies of human belief. It was not that animism was no longer viable—to the contrary, it still seemed to be the most probable explanation for why people believed in spirits. Its disciplinary foothold, therefore, was more or less assured. The main reason for British anthropology's ideological shift in the early twentieth century had to do with researchers no longer viewing spiritual belief as the core of human culture. Since the turn of the century, British anthropologists had worked to complicate the sociocultural landscape, and functionalism reigned supreme among the younger generation of researchers. Bronisław Malinowski (1884–1942) became the discipline's new figurehead, and his fieldwork-centered approach, which employed participant observation as its backbone, came to represent the cutting edge of the research field.[130] Young anthropologists in the early stages of their careers no longer built their work on the research programs of Victorian animists such as Tylor, Lang, and Clodd, whose anthropological models were increasingly seen as old-fashioned armchair pursuits. The exciting research happened abroad in supposed "exotic lands," not in dusty parlor rooms. Studying the cultural history of spiritual belief in the way that Tylor and company had done in their books and articles was the epitome of antiquated Victorian anthropology.[131] With a more sophisticated disciplinary foundation in place, the reality of the spirit hypothesis could never refute anthropology's enriched cultural paradigm.

Outside of anthropology's rapidly changing disciplinary landscape, however, the spiritualist movement continued to flourish. The abominable loss of life resulting from the First World War provided a steady stream of new believers to the spiritualist movement, hoping to make contact with their fallen loved ones. Modern spiritualism experienced a sort of renaissance right through until the end of the Second World War.[132] That is not to say that there were no longer any debates over whether the spirit hypothesis was true. This intellectual battle waged on. Nevertheless, the onus for disproving belief in spirits and psychics increasingly fell to psychologists and stage magicians—some of whom were informed by animistic theories. Ironically, these magicians were the very type of investigator that Lang had argued would be an ideal researcher to study spirit and psychic phenomena.[133] The efforts of figures such as Clodd, however, were not altogether forgotten, and his lasting impact was that he articulated an effective approach for debunking spiritualists that was accessible to all types of investigators from any disciplinary background. His writings, therefore, continued to inform skeptical views long after his death in 1930, among a much broader popular audience.[134]

As a hardline disbeliever, rationalist, and devout follower of Tylorian animism, Clodd believed that spirit and psychic forces were products of either superstition, delusion (caused through illness and intoxication), or trickery. Every so-called genuine case of spirit and psychic phenomena, he argued, was nothing more than misinterpretations of events. A vigilant investigator could study any case and find the causes of error that led to these misperceptions. It did not matter if every single instance of spiritualism or telepathy were exposed as false; all that mattered was that enough doubt was created to question the integrity of the spirit hypothesis overall. Any gaps in the history of exposure did not in itself undermine the merit of a skeptic's argument, so long as plausible doubt remained. With the evidence laid before him, Clodd believed that modern spiritualism had been rigorously tested, and found wanting. The evidence did not support the reality of some hitherto unknown force, but a strong case for the rejection of mediumship as a survival and revival of primitive superstition.

Epilogue

Legacies of Late Victorian Spirit Investigations

In the opening decades of the twentieth century, the famed American magician Harry Houdini (1874–1926) was one of the most ardent investigators of spiritualism on either side of the Atlantic Ocean.[1] He had been fascinated by the movement since the early days of his career, when he was performing as a "mystical entertainer," regularly holding séances for paying sitters.[2] The money was good, but he was highly conscious that he was deceiving people for financial gain. It was not until the death of his mother, Cecilia Steiner (1841–1913), with whom Houdini had a particularly close relationship, that he came to realize the seriousness of his actions: he was preying on vulnerable people, who were desperately trying to have direct contact with deceased loved ones. He wrote, "As I advanced to riper years of experience I was brought to a realization of the seriousness of trifling with the hallowed reverence which the average human being bestows on the departed, and when I personally became afflicted with similar grief I was chagrined that I should ever have been guilty of such frivolity and for the first time realized that it bordered on crime."[3] From that day forward, Houdini began using his first-rate skills as a top stage magician to expose how impostors posing as psychics, produced their extraordinary phenomena during performances.

Houdini, however, suffered from an internal tension. The possibility that there was a genuine form of existence beyond death lingered in his mind. As much as he wanted to expose the swindlers for their abhorrent activities, he also wanted to determine whether there was any real weight to the spirit hypothesis. The prospect of once again communicating with his beloved mother was a key factor in spurring on his psychic investigations. He wrote, "I too would have parted gladly with a large share of my earthly possessions for the solace of one word from my loved departed."[4] There was much at stake emotionally for Houdini, but he still wanted to be critical of the performers he investigated. His desire to communicate with his deceased mother could not outweigh a sagacious inquiry, and he aimed to balance his skepticism and rationalism as a professional magician with his personal desires for potentially verifying a spiritual afterlife. Thus, modern mediumship was to be given a judicious trial before receiving a final verdict from him. He stated that he "never entered a séance room except with an open mind devoutly anxious to learn if intercommunication is within the range of possibilities and with a willingness to accept any demonstration which proves a revelation of truth."[5] Time and time again, however, he found that the mediums he observed used all sorts of deceptive tricks to fool sitters. Every spirit and psychic feat that he witnessed was nothing more than mere illusion.[6]

It might seem strange for a book on late Victorian anthropology's engagement with modern spiritualism to conclude with a discussion on Houdini as a spirit investigator, but in many respects his investigations of alleged spirit and psychic phenomena were a kind of anthropological activity. That is not to say that other modes of inquiry were absent from his investigations, only that his interpretation of spiritualist performances was informed by methods and theories drawn from late Victorian figures such as Alfred Russel Wallace, Edward Burnett Tylor, Andrew Lang, and Edward Clodd. These anthropological influences are not necessarily explicit in Houdini's investigatory work, but through careful scrutiny of his visual epistemology it is possible to thresh them out.[7] An underlying tone of animistic theory informed much of Houdini's views on human belief in spirits and psychics. In some respects, Houdini had taken on the very role as a leading psychic investigator that Lang had argued in 1895 was best suited for "professional experts in conjuring."[8] As the twentieth century pushed forward, investigators such as Houdini, and not professional anthropologists working in universities, took up the mantle of animism in spirit investigations. Those

academics were increasingly focused on the functionalist model of Bronisław Malinowski. For them, animism was a given truth, and no amount
of evidence put before them by spiritualists could convince them otherwise.[9]
However, with the onset of World War I, and a spike in the number of
proponents of the movement, Houdini carried on fighting the old cause of
late Victorian anthropologists.

At the core of Houdini's book *A Magician among the Spirits* (1924) is a pre-
sumption that the continued presence of spiritualist belief in modern society
was a superstitious relic of bygone ages. He wrote, "The ancients' childish
belief in demonology and witchcraft; the superstitions of the civilized and
uncivilized, and those marvelous Mysteries of the past ages are all laughed at
by the full-grown sense of the present generation."[10] Houdini was forwarding
an evolutionary model that underscored how old beliefs merge into new ones,
while often carrying the cultural baggage of earlier societies. However, under
the critical gaze of rationalism, these miraculous marvels could be explained
through naturalistic or mechanical causes. His argument was thus a classic
example of animistic theorizing. That is not to say that Houdini viewed the
world as totally materialistic. He was quite open about his theistic views,
and a higher power was central to his ontological beliefs.[11] Therefore, his
engagement with spiritualism and psychism embodied a mixture of ideas.
There were elements of Wallace's willingness to accept spirits and psychics
as real; Tylor's skepticism and commitment to prima facie evidence; Lang's
flexibility and openmindedness to a potential unseen human agency; and
Clodd's rationalism, materialism, and firm commitment to exposing fraud.

There were other ways in which Houdini's spirit investigations intersected with the research programs of our four main figures. In 1919 alone
he had participated in over one hundred séances in Britain and France, and
he conferred with some of the most prominent spiritualists in Europe and
North America. He even collected an enormous number of books and periodicals on the subject dating as far back as the 1480s, giving him access to
one of the most complete libraries on spiritualism in the world.[12] With rich
resources such as these at his fingertips, he could undertake a meticulous
study of spirits and psychics, drawing on over five hundred years of personal
testimonies. Houdini learned to scrutinize the details of witness accounts
rigorously, identifying patterns, strengths, and flaws. It was a type of critical
reading practice similar to the methods used by Wallace, Tylor, Lang, and
Clodd in their anthropological studies.

Houdini also positioned himself as the epitome of a credible witness: open-minded yet critical, knowledgeable of the secondhand sources but experienced in the séance room. His work was historically informed using comparative methods, while he was also receptive to distinct cultural contexts. Very little separated his research from the works of Wallace, Tylor, Lang, and Clodd. Much like his predecessors, he could take no piece of evidence for granted as legitimate. Even when scrutinizing sources that were recorded by scientifically trained observers, he did not assume that the accounts were credible. Scientific knowledge did not in itself imbue someone with special skills in observing extraordinary phenomena, and "good" witnessing was never a given. Houdini argued that "the fact that they are *scientists* does not endow them with an especial gift for detecting the particular sort of fraud used by mediums, nor does it bar them from being deceived."[13] There were several examples of high-profile scientific researchers supporting the spirit hypothesis, and they exemplified how easily a supposed credible mind could be fooled.[14] Distrusting his sources was just another example of how Houdini's research methods were similar to those of Wallace, Tylor, Lang, and Clodd.

According to Houdini, no two figures were more representative of the folly of scientific minds than Oliver Lodge and the physician and author Arthur Conan Doyle (1859–1930). In the opening decades of the twentieth century, both men published bestselling books that did much to strengthen and foster the spiritualist cause. Lodge's book *Raymond, or, Life after Death* (1916), was a detailed account of his supposed communications with his deceased son. Doyle's book *The New Revelation* (1918) was a testament to the veracity of the spirit hypothesis, anchored in his years of study and direct observation.[15] For Houdini, however, the reason for their misguided views was that they were both "fortified in their belief by grief," which clouded their minds and made them more easily swayed by the so-called positive evidence put before them. Each man had lost a child during the Great War, and was desperate to prove the spirit hypothesis true.[16]

Much like those we have seen in the previous chapters, Houdini argued that most cases in which observers attested to seeing genuine spirit or psychic phenomena were the result of credulity, mental illness, or superstition. The senses are easily fooled, and if one enters a situation holding predetermined expectations, any unusual occurrence can be interpreted as a way of confirming a pre-existing bias. This was particularly the case during the opening decades of the twentieth century, when mourners flocked to séances in the hope of speaking with their dead loved ones.[17] There was an intense desire and willingness for these sitters to accept the phenomena they witnessed as being supernatural in origin. This willingness to accept the feats of psychics as real offered them immense comfort. As Houdini wrote, "distressed relatives catch at the least word which may remotely indicate that the Spirits which they seek is in communication with them. One little sign even, which appeals to their waiting imagination, shatters all ordinary caution and they are converted. Then they begin to accept all kinds of natural events as results of Spirit intervention."[18] Thus, a significant amount of belief in spirits and psychics was the result of mental distress, and this was not so dissimilar to the views of figures such as Tylor, Lang, and Clodd.

What, then, made Houdini a more credible witness and evaluator of spirit and psychic forces? The answer, at least in his view, was his training and experience as a stage magician. His career fundamentally rested on his ability to fool people's perceptions. He wrote, "It has been my life work to invent and publicly present problems, the secrets of which not even the members of the magical profession have been able to discover, and the effects of which have proved as inexplicable to the scientists as any marvel of the mediums."[19] Inventing these incredible illusions helped him develop a visual epistemology and embodied practice that enabled him to determine how other showmen produced their tricks.[20] It was a skill that he employed during his spirit investigations, allowing him to look beyond the surface of a performance, and identify things that other investigators would never have perceived. He boasted that "in so far as the revelation of trickery is concerned my years of investigation have been more productive than the same period of similar work by any scientist."[21] Thus, Houdini possessed a rare skillset, which, when combined with his historical and scholarly knowledge, placed him as a leading spirit and psychic investigator. Unlike most of his predecessors, he could combine his practical grounding in magic with his theoretical knowledge of spiritualism broadly understood.

Some Historiographical Reflections and Lessons Learned

At a micro level, this story of late Victorian anthropology's engagement with modern spiritualism has significant implications for our understanding of the disciplinary history of British anthropology. Although the theory of animism has received some scholarly attention, there has never been a rigorous and systematic historical study of its emergence in the late Victorian era.[22] Much of why its history has been overlooked is the lasting influence of anthropologists such as Alfred Cort Haddon (1885–1940) and Malinowski in the opening decades of the twentieth century. Both men denounced the value of Victorian anthropology as little more than an amateur pursuit, lacking any sort of intellectual muscle.[23] Haddon, for instance, famously argued that Victorian researchers were "retailing second-hand goods over the counter."[24] However, this rhetoric to delegitimize Victorian anthropology was part of the intellectual ecology of the day, which embodied the professionalizing strategies of the new generation of anthropologists, desperate to solidify their positions within the academy.[25]

At a practical level animism still held intellectual merit, even if it had been pushed to the recesses of the discipline. Anthropologists such as Malinowski were trying to supplant evolutionism with functionalism, and animism was caught in the theoretical crossfire, due to its intellectual entanglement with evolutionary thought. Yet, as we have seen in the previous chapters, a revisionist model did not necessitate a developmental framework, and animism could still function as a way of understanding human belief in spirits. The continued appearance of animistic thought in anthropological literature is testament to that.[26] Such a rich historical narrative, however, only becomes clear by re-examining late Victorian anthropology's engagement with modern spiritualism. It was through this interaction that the theory was able to gestate. After all, anthropologists such as Tylor entered séance rooms because of the implications that verified spiritual entities would have on the theoretical limits of animistic thinking. If it were the case that so-called primitive peoples were genuinely observing spirit phenomena in bygone ages, arguments proposing that modern spiritualists were credulous fools and victims of older superstitious thought were wrong. Genuine communications with the spirit world would mean that animism was a false doctrine.

One could also argue that anthropology's engagement with modern spiritualism was a preview of things to come in the discipline. Wallace, Tylor, and Lang all acknowledged the importance of observing spirit and psychic phenomena firsthand. Their spirit investigations, therefore, can be seen as early examples of anthropological fieldwork. Each of these researchers were highly attuned to the limitations of their armchair ruminations, and only by going into the field and testing their ideas could anthropological theories be confirmed. Shifting disciplinary preoccupations were therefore already under way during the height of the so-called armchair era of Victorian anthropology, and spirit investigations are a key example.[27] Even so, the armchair reflections that were being done were not the passive and unsophisticated ruminations that figures such as Haddon had suggested. Rather, they were theoretically nuanced and complex. They engaged directly with spiritualist communities, who often shaped the findings and directed new research programs. The sheer range of animistic interpretations between Tylor, Lang, and Clodd alone highlights this point.[28]

At a macro level, this story further complicates our historical understanding of the relationship between science and religion. Significant historical emphasis has been placed on the idea that Victorians were suffering from a kind of crisis of faith as a result of the cultural transformations occurring during the late Victorian period. However, such a framework generates a problematic understanding of the historical relationship of science and religion.[29] Science and religion did not work in isolation from other cultural factors such as politics, social class, or the law. More often than not an abundance of cultural factors shaped people's psyches, and emphasizing a clash between scientific and religious ideologies alone is overly simplistic as a means of explaining this intellectual unrest. The very act of naturalizing or supernaturalizing human belief is an instance of science and religion coming into contact. Rather than being simply an instance of conflict, as some might expect, it depicts a wonderfully complicated picture.[30]

Nevertheless, as this book has shown, discussions about the existence of spirit and psychic forces often began with questions about belief, but inevitably became immersed in discussions about what constituted reliable evidence. Therefore, a study of anthropology's engagement with modern spiritualism exposes not a story of the so-called Victorian crisis of faith but one of the Victorian crisis of evidence. Our four figures discussed in this book were products of a dramatically changing Victorian world. Darwinism, energy physics, and secularization redefined the intellectual landscape of the nineteenth century, and posed many difficult questions about self, personhood, and existence.[31] It was a period of immense doubt and cognitive dissonance.[32] Old taken-for-granted assumptions about the place of humans in the world were fading, and movements such as modern spiritualism, and theories such as animism, were products of this shifting cultural paradigm.

The observational practices of Wallace, Tylor, Lang, and Clodd demonstrate how their responses to these ideas affected their interpretations of spiritualism and extraordinary belief in different ways, and show a range of reactions and perspectives to these cultural pressures in Victorian society.[33]

Wallace, for example, sought to align his naturalistic and materialistic understanding of evolutionary processes with his immaterial and ethereal notions of spirits and psychics. The result was what he viewed as a supernormal existence, where modern spiritualism represented the culmination of scientific and spiritual ideas merging together to form the "true religion."[34] He conceptualized this process as forming part of his "theory of spiritualism," combining "natural selection" with "progression of the fittest."[35] Tylor's response to these cultural pressures was different. He was more representative of an open-minded skepticism from the period. He did not reject spiritualist claims outright, but he was unwilling to accept the spirit hypothesis as true until he had observed credible prima facie evidence of probable spirits or psychic forces. His trip to London in November of 1872 was testament to his readiness to give modern spiritualism a fair hearing, but his rationalism and naturalism were unmoved by the performances that he witnessed. That is not to say that he settled on a firm rejection of spiritualism; he maintained that he would continue to examine new evidence that might alter his position.[36]

Lang's engagement with modern spiritualism represents a different response to the changing cultural views of the Victorian age. His initial rejection of modern spiritualism was the result of his early commitment to animistic theory. However, as he matured intellectually, he became increasingly sympathetic to the spiritualist cause. He did not become a supporter of modern spiritualism in the same way as Wallace, but instead he became a proponent of telepathy, another example of naturalistic and spiritualistic ideas merging. If telepathy was the product of some unseen human faculty, it was natural in origin, but with extraordinary, immaterial implications. Thus, Lang's support of telepathy still resided within the broad remit of scientific naturalism, albeit one that was different from both Wallace's and Tylor's visions of the paradigm.[37] Clodd, on the other hand, was highly antagonistic toward religious and spiritualistic ideas from the onset. His two books *The Childhood of the World* and *The Childhood of Religions* are indicative of a larger naturalistic and rationalistic opposition to modern spiritualism and telepathy during the period. Clodd was not an open-minded skeptic like Tylor; instead, he was a hardline disbeliever of all supernatural or supernormal phenomena. He embodied much of the secularist and materialist mindset emerging throughout the second half of the nineteenth century.[38]

In all four of the case studies from the preceding chapters, theories such the spirit hypothesis and animism may have arisen as points of intellectual contestation between different philosophical camps, but the debates that ensued were inevitably very practical, and relied heavily on experimental and experiential testing and data analysis. What counted as reliable evidence for spirit investigators, therefore, fed into larger discussions of the standards of evidence in Victorian science more broadly. Ideas and beliefs do not function in a separate sphere, and always have an evidentiary element to them. Through a critical consideration of the evidence-based foundations of Victorian spirit investigations, we come to better understand the kinds of philosophical assumptions that nineteenth-century researchers brought to bear on the subject of spiritualism as they struggled to interpret different sets of data that did not conform to dominant modes of knowledge, such as scientific naturalism and rationalism. The limits of these theories, and the evidence used to support them, was repeatedly tested, requiring investigators to regularly reassess and strengthen their practices.

Finally, this story of late Victorian anthropology's engagement with modern spiritualism provides an important and nuanced narrative on the history of visual epistemologies in nineteenth-century human sciences.[39] A core question throughout the book is: who should be trusted as a credible witness of spirit and psychic phenomena? Understanding what made for trustworthy observation was a key preoccupation for all of the historical actors in each of the four case studies.[40] As the chapters have shown, these researchers used a variety of verification methods for legitimizing observations, including collective testimony, replication, and virtual recreation through text, to name a few key examples.[41] Because anthropology in particular relied so heavily on experiential knowledge as the foundation of its praxis, it was important for researchers to develop highly sophisticated methods for assessing the veracity of personal testimonies. Witness accounts were especially susceptible to all sorts of personal biases, and therefore it was essential for anthropologists using them in their investigations to be able to develop strong cases for establishing their credibility as key sources of evidence. Thus, the process of turning any old observer into a credible witness was rigorous and methodical. A researcher's reputation as a trustworthy scientific investigator was based on his or her ability to either confirm the earlier observations of witnesses who had seen a particular type of cultural phenomenon, or correct them based on newer information. Direct experience was essential to this process, and knowledge based on prima facie evidence added greatly to a researcher's truth-claims.[42]

Ultimately, the spirit investigations of Wallace, Tylor, Lang, and Clodd show that there was a long history of anthropologists transforming personal testimonies into scientific evidence, but because of the extraordinary nature of spirit and psychic forces, it was even more important for researchers to establish the accuracy and authenticity of their informants.[43] Such historiographical reflections have important implications for our understanding of what constitutes "proper" scientific practice. While scientific engagement with modern spiritualism has typically been seen as an amusing side story in the history of science, it was actually quite important to the actors themselves. So many high-profile scientific figures engaged with modern spiritualism because of its major implications for the foundation of modern naturalistic science.[44] The reality of an immaterial existence outside the boundaries of a natural and material world cut to the core of scientific naturalism. A study of its history, therefore, is a study of the making of the modern world.

NOTES

Introduction

1. E. B. Tylor to A. R. Wallace, November 26, 1866, Add 46439 ff. 6, Tylor Papers, British Library.

2. The most notable example is Oppenheim, *Other World*.

3. For more on rationalistic doubt in the Victorian age see Larsen, *Crisis of Doubt*; Lane, *Age of Doubt*; and Franklin, *Spirit Matters*. For more on cognitive dissonance see Moore, *Post-Darwinian Controversies*, 14, 111–13, and 121; Festinger, *Theory of Cognitive Dissonance*; Stocking, *After Tylor*, 57–59; and Oppenheim, *Other World*, 1–3.

4. Thomson, "On a Universal Tendency in Nature," 304–6. Lord Kelvin further elaborated on his concept of solar entropy in Thomson, "On the Age of the Sun's Heat," 388–93. See also Gilmour, *Victorian Period*, 36–37; Smith, *Science of Energy*, 53, 142, 314; Jones, "Gone into Mourning," 178–95; Alexander, *Victorian Literature and the Physics of the Imponderable*, 51–82; Krage, *Entropic Creation*, 23–46.

5. For a more nuanced understanding of the boundaries between science and religion see Turner, "Victorian Conflict between Science and Religion," 356–76; Lightman, *Origins of Agnosticism*; Brooke and Cantor, "Whose Science? Whose Religion?" 43–72; Bowler, *Reconciling Science and Religion*; Harrison, "'Science' and 'Religion,'" 81–106; Lightman, "Victorian Sciences and Religion," 343–66; Lightman, "Does the History of Science and Religion Change," 149–68; Brooke, *Science and Religion*; Stanley, *Huxley's Church and Maxwell's Demon*; and Harrison, *Territories of Science and Religion*.

6. Much has been written in the secondary literature on the staggering amount of scientific, medical, and technological transformation occurring during the nineteenth century. Some key examples include Headrick, *Tools of Empire*; Adas, *Machines as the Measure of Men*; Diamond, *Guns, Germs, and Steel*; Marsden and Smith, *Engineering Empires*; and Headrick, *Power over Peoples*.

7. The first use of the term *psychical research* to denote the scientific study of psychics or mediums appeared in Gurney, Podmore, and Myers, *Phantasms of the Living*, 1:5.

8. Oppenheim, *Other World*, 136–38.

9. For more on the theory of animism see Stocking, "'Cultural Darwinism' and 'Philosophical Idealism,'" 91–109; Stocking, "Animism in Theory and Practice," 88–104; Leopold, *Culture in Comparative and Evolutionary Perspective*; Stringer, "Rethinking Animism," 541–55; Strenski, *Thinking about Religion*, 91–116; Ratnapalan, "E. B. Tylor and the Problem of Primitive Culture," 131–42; Sera-Shriar, *Making of British Anthropology*, 147–76; and Sera-Shriar, "Historicizing Belief," 68–90.

10. Stocking, "Animism in Theory and Practice," 100.

11. For more on the crisis of evidence in spirit and psychic investigations see Lamont, "Spiritualism and a Mid-Victorian Crisis of Evidence," 897–920; and Noakes, "Spiritualism, Science and the Supernatural," 24–25.

12. I am borrowing the term *visual epistemologies* from Bleichmar, *Visible Empire*, 6–10.

13. For more on "visual epistemologies" and observational study in Victorian anthropology see Sera-Shriar, *Making of British Anthropology*, 1–20; and Sera-Shriar, "What Is Armchair Anthropology?" 180–94. For more secondary literature on the history of scientific observation within the natural and social sciences see Coon, "Testing the Limits of Sense and Science," 143–51; Secord, "Artisan Naturalists," 135–206; Grimshaw, *Ethnographer's Eye*; Grasseni, *Skilled Visions*; Daston and Galison, *Objectivity*; and Daston and Lunbeck, *Histories of Scientific Observation*. See also Tedlock, "From Participant Observation to the Observation of Participation," 61–94; Stocking, *Ethnographer's Magic and Other Essays*; Gupta and Ferguson, "Discipline and Practice," 1–46; Kuklick, "After Ishmael," 47–65; and Kuklick, "Personal Equations," 1–33.

14. Sera-Shriar, *Making of British Anthropology*, 1–20.

15. For more on colonial exchange networks in botany see Secord, "Corresponding Interests," 383–408; Endersby, *Imperial Nature*; Laidlaw, *Colonial Connections, 1815–1845*; Bleichmar, *Visible Empire*.

16. For more on the use of instructive literature for informants within British ethnology and anthropology see Bravo, "Ethnological Encounters," 338–57; and Urry, "Notes and Queries on Anthropology," 45–57; Sera-Shriar, *Making of British Anthropology*, 53–79, 147–76.

17. For more on truth and testimony see Shapin, *Social History of Truth*; Hacking, *Social Construction of What?*; Golan, *Laws of Men and Laws of Nature*; Sera-Shriar, "Arctic Observers," 23–31; Sera-Shriar, "Tales from Patagonia," 204–23; Baggini, *Short History of Truth*; and Kaalund, "From Science in the Arctic to Arctic Science."

18. For more on "virtue epistemics" see Daston and Galison, *Objectivity*, 39–42.

19. There have been two recent studies on the intersection of Alfred Russel Wallace's anthropology and spiritualism, but these are by no means a comprehensive disciplinary picture. See Ferguson, "Other Worlds," 177–91; and Sera-Shriar, "Credible Witnessing," 1–28.

20. Luckhurst, *Invention of Telepathy*; Larsen, *Slain God*; Richardson, *Second Sight in the Nineteenth Century*; Josephson-Storm, *Myth of Disenchantment*; Raia, *New Prometheans*; and Ferguson, *Determined Spirits*.

21. Oppenheim, *Other World*; Noakes, "Spiritualism, Science and the Supernatural," 23–43; Noakes, "Haunted Thoughts of the Careful Experimentalist," 46–56; McCorristine, *Spectres of the Self*; Lamont, *Extraordinary Beliefs*. See also Mauskopf and McVaugh, *Elusive Science*; Collins and Pinch, *Frames of Meaning*; Roach, *Spook*; Blum, *Ghost Hunters*; and Asprem, *Problem of Disenchantment*.

22. Barrow, *Independent Spirits*.

23. Owen, *Darkened Room*.

24. Cottom, *Abyss of Reason*; and Sword, *Ghostwriting Modernism*.

25. Shapin and Schaffer, *Leviathan and the Air-Pump*; Golan, *Laws of Men and Laws of Nature*; Daston and Galison, *Objectivity*.

26. Shapin and Schaffer, *Leviathan and the Air-Pump*, 55–65.

27. Chadwick, *Victorian Church*; Pals, *Victorian "Lives" of Jesus*; Shea and Whitla, *Essays and Reviews*; Hesketh, *Victorian Jesus*. See also Bennett, *God and Progress*.

28. Luckhurst, *Invention of Telepathy*, 1.

29. For more on historicism as an interpretive tool see Stocking, "On the Limits of 'Presentism' and 'Historicism,'" 1–12; Kuklick, "Introduction," 1–16; Sera-Shriar, "From the Beginning," 1–13.

30. For more on the relationship between the "observer" and the "observed" in the history of the human sciences see Stocking, *Observers Observed*; and Kuklick, "Personal Equations," 1–133.

31. Hall, *Spiritualists*; Hall, *Strange Case of Edmund Gurney*; and Hall, *Strange Case of Ada Goodrich Freer*.

32. Gauld, *Founders of Psychical Research*.

33. Oppenheim, *Other World*; Brandon, *Spiritualists*; and Cerullo, *Secularisation of the Soul*.

34. Noakes, *Physics and Psychics*, 8–9.

35. Examples of important works on Victorian anthropology are Stocking, *Victorian Anthropology*; Kuklick, *Savage Within*; Kuklick, *New History of Anthropology*; Qureshi, *Peoples on Parade*; Manias, *Race, Science and the Nation*; Sera-Shriar, *Making of British Anthropology*; Flandreau, *Anthropologists in the Stock Exchange*.

36. Stocking, *Ethnographer's Magic and Other Essays*, 3–4.

37. Ginzburg, *Cheese and the Worms*; Davis, *Return of Martin Guerre*; and Magnússon and Szjártó, *What Is Microhistory?*. See also Ginzburg, "Microhistory," 10–35; Levi, "On Microhistory," 93–113; Tristano, "Microhistory and Holy Family Parish," 23–30; Magnússon, "'Singularization of History,'" 701–35; Brown, "Microhistory and the Post-Modern Challenge," 1–20; Magnússon, "Social History as 'Sites of Memory'?" 891–913; Magnússon, "Microhistory, Biography and Ego-Documents in Historical Writing," 133–53; Magnússon, "Far-Reaching Microhistory," 312–41.

38. Geertz, *Interpretation of Cultures*, 25–26.

39. Geertz, *Interpretation of Cultures*, 5.

40. Wallace, *On Miracles and Modern Spiritualism*.

41. Alfred Russel Wallace to Thomas Henry Huxley, November 22, 1866, Add. 46439 f. 5, Wallace Papers, British Library, London.

42. Edward Burnett Tylor, *Notebook on Spiritualism*, item 12, box 3, Pitt Rivers Museum Manuscript Collection, University of Oxford.

43. Tylor, *Primitive Culture*.

44. Clodd, *Question*.

Chapter 1: Alfred Russel Wallace

1. For more details on Wallace's first séance experiences at Leslie's home see Alfred Russel Wallace, Journal, July 22, 1865, pp. 39–42, WCP5223.5749, Papers of Alfred Russel Wallace, Natural History Museum, London. Wallace had been courting Leslie's daughter Marian (b. 1846) in 1864, but she ended their relationship in January 1865. Nevertheless, Wallace and the Leslie family remained good friends. For more, see Wyhe, *Dispelling the Darkness*, 310–11.

2. Wallace, *On Miracles and Modern Spiritualism*, vi–vii.

3. Wallace defined modern spiritualism as the practice of the living communicating with the dead through psychic mediums. See Wallace, *On Miracles and Modern Spiritualism*, 80.

4. Fichman, "Science in Theistic Contexts," 231–32.

5. Alfred Russel Wallace to Thomas Henry Huxley, November 22, 1866, BL Add. 46439 f. 5, British Library, London.

6. Marchant, *Alfred Russel Wallace*, 2:187–88.

7. For more on the debates over reliable evidence in spirit investigations see Lamont, "Spiritualism and a Mid-Victorian Crisis of Evidence," 897–920.

8. Oppenheim, *Other World*, 297.

9. For more on the establishment of credible witnesses in ethnology and anthropology see Sera-Shriar, "Arctic Observers," 23–31; and Sera-Shriar, "Tales from Patagonia," 204–23.

10. For more on observational practices within Victorian anthropology see Sera-Shriar, "What Is Armchair Anthropology?" 180–94.

11. Bleichmar, *Visible Empire*, 6–10. For more secondary literature on the history of scientific observation within the natural and social sciences see Secord, "Artisan Naturalists," 135–206; Grasseni, *Skilled Visions*; Grimshaw, *Ethnographer's Eye*; and Daston and Galison, *Objectivity*.

12. See, for example, the following works on Wallace's spirit investigations: Keezer, "Alfred Russel Wallace," 66–70; Kottler, "Alfred Russel Wallace, the Origin of Man, and Spiritualism," 144–92; Durant, "Scientific Naturalism and Social Reform," 31–58; Pels, "Spiritual Facts and Super-Visions," 69–91; Fichman, "Science in Theistic Contexts," 227–50; and Mitchell, "Capturing the Will," 15–24. Wallace's spiritualism has also been discussed in the following biographies: Shermer, *In Darwin's Shadow*; Slotten, *Heretic in Darwin's Court*, 326–51; and Fichman, *Elusive Victorian*, 139–208.

13. The best example of a scholar to investigate Wallace's interest in extra-European spiritualism is Ferguson, "Other Worlds," 177–91. For more on the intersection of Wallace, anthropology, travel, and spiritualism see Lowrey, "Alfred Russel Wallace as Ancestor Figure," 18–21.

14. James Hunt discussed the importance of Baconianism in his article "On Physio-Anthropology, Its Aim and Method," ccix–cclxxi. For more on Hunt's use of Baconianism in anthropology see Sera-Shriar, "Observing Human Difference," 480–85.

15. For more on Wallace's activities at the Anthropological Society of London, see Vetter, "Unmaking of an Anthropologist," 25–42.

16. Evelleen Richards was the first scholar to frame Hunt and Huxley's anthropological feud as a competition between two forms of scientific naturalism in her article "'Moral Anatomy' of Robert Knox," 373–436. For more on the contest for cultural authority in the nineteenth century see Turner, "Victorian Conflict between Science and Religion," 356–76; and Turner, *Contesting Cultural Authority*.

17. For more on the disciplinary debates between British anthropology and ethnology during the 1860s see Stocking, "What's in a Name?" 369–90; Stocking, *Victorian Anthropology*, 238–45; Kenny, "From the Curse of Ham to the Curse of Nature," 367–88; Kuklick, "British Tradition," 52–56; and Sera-Shriar, *Making of British Anthropology*, 109–46. For more on evolutionism and historicism in the human sciences see Sera-Shriar, "Human History and Deep Time in Nineteenth-Century British Sciences," 19–22; and Sera-Shriar, *Historicizing Humans*.

18. Hunt and Huxley outlined the parameters of anthropology and ethnology respectively in the following essays: Hunt, "Introductory Address on the Study of Anthropology," 1–20; Huxley, "On the Methods and Results of Ethnology," 257–77.

19. For more see Stocking, "What's in a Name?" 369–90; and Sera-Shriar, "Race," 48–49.

20. Hunt, "Introductory Address," 2.

21. Durant, "Scientific Naturalism and Social Reform," 33.

22. Wilson, *Forgotten Naturalist*, 6–10. See also Jones, "Alfred Russell Wallace, Robert Owen and the Theory of Natural Selection," 73–96.

23. Owen, *Future of the Human Race*. For more on Owen's conversion to spiritualism see Podmore, *Robert Owen*, 2:604–5. See also Oppenheim, *Other World*, 11, 273, and 290; Owen, *Darkened Room*, 19; and Wiley, *Thought Reader Craze*, 9–17.

24. Wallace, *On Miracles and Modern Spiritualism*, 109. See also Pels, "Spiritual Facts and Super-Visions," 74–77.

25. Slotten, *Heretic in Darwin's Court*, 10–16; and Wyhe, "Introduction," 3–5.

26. Alfred Russel Wallace to Thomas Henry Huxley, November 22, 1866, BL Add. 46439 f. 5, British Library, London.

27. For more on the disciplinary practices of British anthropology in the middle of the nineteenth century see Sera-Shriar, "Observing Human Difference," 461–91.

28. Wallace, *On Miracles and Modern Spiritualism*, 125–26.

29. Wallace, *My Life*, 1:217.

30. Wallace, *My Life*, 1:218.

31. For more on Spencer Timothy Hall and mesmerism see Winter, *Mesmerized*, 130–36.

32. Wallace, *On Miracles and Modern Spiritualism*, 119.

33. Fichman, "Science in Theistic Contexts," 231.

34. Wallace, *On Miracles and Modern Spiritualism*, 125.

35. For Wallace's theory of spiritualism see Wallace, *On Miracles and Modern Spiritualism*, 100–104. In addition to mesmerism, there were other ideas shaping Wallace's views on spiritualism. For example, although he does not engage with the writings of the Swedish theologian, philosopher, and mystic Emanuel Swedenborg (1688–1772) in *On Miracles and Modern Spiritualism*, Swedenborgian ideas came to be important for Wallace's later writings on spirits and psychic forces. For more on this connection see Fichman, *Elusive Victorian*, 112–17.

36. Weisberg, *Talking to the Dead*, 12–13.

37. For more on Cora L. V. Scott see Barrett, *Life of Mrs. Cora L.V. Richmond*.

38. Oppenheim, *Other World*, 11, 273, and 290; Owen, *Darkened Room*, 19; Wiley, *Thought Reader Craze*, 9–17.

39. It is estimated that by the end of the nineteenth century there were around eight million spiritualists in Europe and North America. See Braude, *Radical Spirits*, 296. For more on the rise of spiritualism in the nineteenth century see Walkowitz, "Science and the Séance," 3–29; Holloway, "Enchanted Spaces," 182–87; Davies, *Haunted*; Bann, "Ghostly Hands and Ghostly Agency," 663–85; McCorristine, *Spectres of the Self*, 12–13; Nartonis, "Rise of Nineteenth-Century American Spiritualism," 361–73; and Sommer, "Psychical Research in the History and Philosophy of Science," 38–45.

40. Fichman, "Science in Theistic Contexts," 231. For more on the link between Wallace's travels, ethnography, and spiritualism see Fichman, *Elusive Victorian*, 170–71; Lyons, *Species, Serpents, Spirits and Skulls*, 120; Ferguson, "Other Worlds," 177–91.

41. For example, the geologist and zoologist Robert Jameson's natural history classes at the University of Edinburgh were designed to prepare young medics and naturalists for careers as scientific travelers. Included in his reading lists were important travel narratives. See Desmond and Moore, *Darwin's Sacred Cause*, 28–29.

42. For more on ethnography and observational practices see Sera-Shriar, "Tales from Patagonia," 206–9.

43. Lyons, *Species, Serpents, Spirits and Skulls*, 120.

44. Bravo, "Ethnological Encounters," 344; and Sera-Shriar, "Arctic Observers," 23–31. For more on the importance of travel literature in the making of natural sciences generally see Hulme and Youngs, "Introduction," 1–16; Browne, "Science of Empire," 453–75; Koerner, "Purposes of Linnaean Travel," 117–52; Carey, "Compiling Nature's History," 269–92; and Kaalund, "From Science in the Arctic to Arctic Science."

45. For more on "truth-to-nature objectivity" see Daston and Galison, *Objectivity*, 55–113.

46. For more on Humboldtian science see Cannon, *Science in Culture*; Pratt, *Imperial Eyes*,

111–44; Carey, "Compiling Nature's History," 269–92; Dettelbach, "Face of Nature," 473–504; and Wulf, *Invention of Nature*.

47. Durant, "Scientific Naturalism and Social Reform," 39. For more on Prichard and Lawrence as founding figures of British ethnology see Sera-Shriar, *Making of British Anthropology*, 21–52. Prichard and Lawrence's foundational works in ethnology are Prichard, *Researches into the Physical History of Man*; and Lawrence, *Lectures on Physiology, Zoology, and the Natural History of Man*.

48. Durant, "Scientific Naturalism and Social Reform," 40; Vetter, "Unmaking of an Anthropologist," 25–42; and Stocking, *Victorian Anthropology*, 148–49.

49. Wallace's original three essays appeared as follows: *Scientific Aspect of the Supernatural*; "Answer to the Arguments of Hume, Lecky, and Others," 113–16; "Defence of Modern Spiritualism: Part I," 630–57; and "Defence of Modern Spiritualism: Part II," 785–807.

50. Wallace, *On Miracles and Modern Spiritualism*, 100. For more on reliable evidence in spiritualism see Noakes, "Haunted Thoughts of the Careful Experimentalist," 46–56; Lamont, "Spiritualism and a Mid-Victorian Crisis of Evidence," 897–920.

51. Wallace, *On Miracles and Modern Spiritualism*, p. 100.

52. Oppenheim, *Other World*, 307; and Fichman, *Elusive Victorian*, 141.

53. For more on scientific naturalism see: MacLeod, "X Club," 305–22; Turner, "Victorian Conflict between Science and Religion," 356–76; Barton, "Huxley, Lubbock, and Half a Dozen Others," 410–44; Desmond, "Redefining the X Axis," 3–50; and Dawson and Lightman, *Victorian Scientific Naturalism*.

54. Wallace, *On Miracles and Modern Spiritualism*, 100.

55. Wallace, *On Miracles and Modern Spiritualism*, 109.

56. For a more nuanced understanding of the boundaries between science and religion see Turner, "Victorian Conflict between Science and Religion," 356–76; Brooke and Cantor, "Whose Science? Whose Religion?" 43–72; Harrison, "'Science' and 'Religion,'" 81–106; Lightman, "Victorian Sciences and Religion," 343–66.

57. For more on evolutionism in Victorian ethnology and anthropology see Stocking, *Race, Culture and Evolution*; Stocking, "'Cultural Darwinism' and 'Philosophical Idealism,'" 91–109; Leopold, *Culture in Comparative and Evolutionary Perspective*; Sera-Shriar, "Observing Human Difference," 468–69; Kuklick, "Theory of Evolution and Cultural Anthropology," 83–102; Sera-Shriar, "Race," 48–76.

58. Wallace, "Origin of Human Races and the Antiquity of Man," clviii–clxxxvii.

59. Although Wallace frames his evolutionary paradigm as Darwinian, his theistic tendencies were incommensurable with Darwin's inherent positivism and Humean predilection. For more on Wallace's evolutionism see Flannery, "Alfred Russel Wallace, Nature's Prophet," 51–70; Fichman, *Elusive Victorian*. For more on Darwin, positivism, and Humean philosophy see Gillespie, *Charles Darwin and the Problem of Creation*; and Brown, "Evolution of Darwin's Theism," 1–45.

60. Wallace, *On Miracles and Modern Spiritualism*, 109.

61. Wallace, *On Miracles and Modern Spiritualism*, 101.

62. For more see Harrison, "'Steam Engine of the New Moral World,'" 76–98; Garnett, *Co-operation and the Owenite Socialist Communities in Britain*.

63. Fichman, "Science in Theistic Contexts," 228. See also Wallace, *On Miracles and Modern Spiritualism*, 116.

64. Wallace, *On Miracles and Modern Spiritualism*, 115–16. See also Pels, "Spiritual Facts and Super-Visions," 74–77.

65. For more on Hunt and the Anthropological Society of London see Burrow, "Evolution and Anthropology in the 1860's [*sic*]," 137–49; Stocking, "What's in a Name?" 369–90; Stocking, *Victorian Anthropology*, 247–55; Spencer, "Hunt, James (1833–1869)," 506–8; Kenny, "From the Curse of Ham to the Curse of Nature," 367–88; Kuklick, "British Tradition," 52–55; and Sera-Shriar, "Observing Human Difference," 461–91.

66. For more on Baconianism and disciplinary formation in nineteenth-century Britain see Rudwick, *Great Devonian Controversy*, 24–25; Porter, *Making of Geology*, 66–70; and Yeo, "Scientific Method and the Rhetoric of Science in Britain," 259–97.

67. Gruber, *Darwin on Man*, 122.

68. Hunt, "On Physio-Anthropology," ccxii.

69. For more on the race taxonomies of Prichard and Lawrence see Prichard, *Researches into the Physical History of Man*, 7, 21–25; Lawrence, *Lectures on Physiology, Zoology, and the Natural History of Man*, 107, 242–52. See also Sera-Shriar, *Making of British Anthropology*, 27–42.

70. Tylor, *Primitive Culture*, 1:vi. See also Sera-Shriar, *Making of British Anthropology*, 163–64.

71. Wallace, *On Miracles and Modern Spiritualism*, 47.

72. Bleichmar, *Visible Empire*, 6–10.

73. Wallace, *On Miracles and Modern Spiritualism*, 32.

74. Wallace, *On Miracles and Modern Spiritualism*, 32.

75. See Shapin and Schaffer, *Leviathan and the Air-Pump*, 55–65; Noakes, "Haunted Thoughts of the Careful Experimentalist," 46–56. For more on measurement, standardization, and science generally see Gooday, *Morals of Measurement*, 1–39.

76. Wallace, *On Miracles and Modern Spiritualism*, 199–200.

77. Wallace, *On Miracles and Modern Spiritualism*, 200–202. For more on the history of trance see Taves, *Fits, Trances, and Visions*.

78. Wallace, *On Miracles and Modern Spiritualism*, 202.

79. Wallace, *On Miracles and Modern Spiritualism*, 104.

80. The use of travel narratives to substantiate ethnological and anthropological research was widespread within the two disciplines, and is sprinkled throughout the works of nineteenth-century researchers. For example, see Prichard, *Researches into the Physical History of Man*; Lawrence, *Lectures on Physiology, Zoology, and the Natural History of Man*; Latham, *Natural History of the Varieties of Man*; Hunt, *On the Negro's Place in Nature*; Tylor, *Primitive Culture*.

81. Vetter, "Unmaking of an Anthropologist," 25–42; Durant, "Scientific Naturalism and Social Reform," 31–58.

82. Daston and Galison, *Objectivity*, 19–27. See also Bleichmar's discussion on long-distance observation in Bleichmar, *Visible Empire*, 66–72.

83. For more on observation and ethnography see Browne, "Biogeography and Empire," 306–8; Bravo, "Ethnological Encounters," 338–57; Sera-Shriar, "Tales from Patagonia," 206–9; Sera-Shriar, "Arctic Observers," 23–31.

84. Shapin and Schaffer, *Leviathan and the Air-Pump*, 55–65.

85. Wallace, *On Miracles and Modern Spiritualism*, 145.

86. For more on how armchair ethnologists made sense of others' firsthand observations see Sera-Shriar, *Making of British Anthropology*, 75.

87. Wallace, *On Miracles and Modern Spiritualism*, 47–48.

88. Wallace, *On Miracles and Modern Spiritualism*, 51.

89. Much has been written on Kate Fox and her sisters. For example, see Davenport, *Death-Blow to Spiritualism*; Oppenheim, *Other World*, 11, 32, 35, 126, 297, 331, 344; Weisberg, *Talking to the Dead*; Chapin, *Exploring Other Worlds*; McCorristine, *Spectres of the Self*, 12, 60, 90.

90. Wallace, *On Miracles and Modern Spiritualism*, 156. It is worth noting, however, that in 1888 Kate and Margaret Fox publicly admitted to committing fraud during one of their performances. See Podmore, *Modern Spiritualism*, 187–88.

91. Owen, *Footfalls on the Boundary of Another World*; and Owen, *Debatable Land between This World and the Next*. For more on Owen and spiritualism see Lehman, *Victorian Women and the Theatre of Trance*, 160–68.

92. Wallace, *On Miracles and Modern Spiritualism*, 157.

93. [Chambers], *Vestiges of the Natural History of Creation*.

94. For more on Chambers and spiritualism see Oppenheim, *Other World*, 272–78.

95. Wallace, *On Miracles and Modern Spiritualism*, 156.

96. Luckhurst, *Invention of Telepathy*, 26.

97. For more on Faraday's experiments see Faraday, "Experimental Investigation of Table-Turning," 328–33; "Michael Faraday's Researches in Spiritualism," 145–50; and Oppenheim, *Other World*, 327–28, 336–37.

98. Carpenter, "Electro-Biology and Mesmerism," 501–57.

99. Lamont, "Spiritualism and a Mid-Victorian Crisis of Evidence," 901–3.

100. The most comprehensive study of Home's career is Lamont, *First Psychic*.

101. Wallace, *On Miracles and Modern Spiritualism*, 158–59. The original source of these observations come from Brewster's daughter's biography of her father: Gordon, *Home Life of Sir David Brewster*, 257–58.

102. Wallace, *On Miracles and Modern Spiritualism*, 158–9. See also Lamont, "Spiritualism and a Mid-Victorian Crisis of Evidence," 901–2.

103. Wallace, *On Miracles and Modern Spiritualism*, 161.

104. Wallace, *On Miracles and Modern Spiritualism*, 161. For the original source see Cox, *What Am I?*, 2:388.

105. For more on Guppy see Richard Noakes, "Guppy, (Agnes) Elisabeth (1838–1917)," *Oxford Dictionary of National Biography*, http://www.oxforddnb.com/view/article/54083.

106. Owen, *Darkened Room*, 42.

107. Wallace's first series of séance observations relating to Guppy are recorded in his journal between November 16 and December 14 of 1866. See Alfred Russel Wallace, Journal, November 23, 1866, to December 14, 1866, pp. 75–78, WCP5223.5749, Papers of Alfred Russel Wallace, Natural History Museum, London.

108. Wallace, *On Miracles and Modern Spiritualism*, 163.

109. Alfred Russel Wallace, Journal, December 7, 1866, p. 75, WCP5223.5749, Papers of Alfred Russel Wallace, Natural History Museum, London.

110. Shapin and Schaffer, *Leviathan and the Air-Pump*, 60–65.

111. Bleichmar, *Visible Empire*, 6–10.

112. Wallace, *On Miracles and Modern Spiritualism*, 126. Wallace recorded a detailed account of this séance in his notebook from the entry for July 22, 1865: Alfred Russel Wallace, Journal, July 22, 1865, pp. 39–42, WCP5223.5749, Papers of Alfred Russel Wallace, Natural History Museum, London.

113. Wallace, *On Miracles and Modern Spiritualism*, 126.

114. Geertz, *Interpretation of Cultures*, 5–6.

115. Alfred Russel Wallace, Journal, July 22, 1865, p. 42, WCP5223.5749, Papers of Alfred Russel Wallace, Natural History Museum, London.

116. Wallace, Journal, July 22, 1865, 42.

117. For more on visual epistemologies see Bleichmar, *Visible Empire*, 6–10.

118. Alfred Russel Wallace, Journal, May 29, 1867, pp. 91–92, WCP5223.5749, Papers of Alfred Russel Wallace, Natural History Museum, London.

119. For more on Guppy's apports see Owen, *Darkened Room*, 42.

120. Alfred Russel Wallace, Journal, November 23, 1866, p. 77, WCP5223.5749, Papers of Alfred Russel Wallace, Natural History Museum, London.

121. Alfred Russel Wallace, Journal, June 21, 1867, p. 98, WCP5223.5749, Papers of Alfred Russel Wallace, Natural History Museum, London.

122. Wallace, Journal, June 21, 1867, p. 98.

123. Sera-Shriar, "Anthropometric Portraiture and Victorian Anthropology," 155–79.

124. Photographs were widely used as credible sources in science during the nineteenth century. For more information see Tucker, "Photography as Witness, Detective, and Impostor," 378–408; Sera-Shriar, "Anthropometric Portraiture and Victorian Anthropology," 165. See also Tucker, "Historian, the Picture and the Archive," 111–20; and Mifflin, "Visual Archives in Perspective," 32–69.

125. Wallace, *On Miracles and Modern Spiritualism*, 186.

126. Wallace, *On Miracles and Modern Spiritualism*, 186.

127. Wallace, *On Miracles and Modern Spiritualism*, 186.

128. Wallace, *On Miracles and Modern Spiritualism*, 186.

129. For work on Frederick Hudson and spirit photography generally see Kaplan, "Where the Paranoid Meets the Paranormal," 18–27; Chéroux, Fischer, Apraxine, Canguilhem, and Schmit, *Perfect Medium*; Jolly, *Faces of the Living Dead*; Tucker, *Nature Exposed*, 98–103; and Harvey, *Photography and Spirit*; Natale, "Short History of Superimposition," 125–45.

130. Wallace, *On Miracles and Modern Spiritualism*, 191.

131. Tylor, *Primitive Culture*, 1:384–85.

132. Alfred Russel Wallace to Thomas Henry Huxley, November 22, 1866, BL Add. 46439 f. 5, British Library, London.

133. Wallace, "Primitive Culture," 69–71.

134. It is worth noting that by 1892 Wallace no longer viewed spiritualism as supernatural, and regretted having formed this link in his 1866 essay. See Wallace, "Spiritualism," 645–49.

135. For more on the formation of the Society for Psychical Research see Bennett, *Society for Psychical Research*; Salter, *Society for Psychical Research*; Gauld, *Founders of Psychical Research*; Haynes, *Society for Psychical Research*; Oppenheim, *Other World*, 142–58, 253–62, and 361–65; and Luckhurst, *Invention of Telepathy*, 51–59, and 148–50.

Chapter 2: Edward Burnett Tylor

1. Buckland, *Spirit Book*, 81; Warner, *Phantasmagoria*, 227; and Robertson, *Science of the Séance*, 142.

2. Oppenheim, *Other World*, 17–21.

3. Owen, *Darkened Room*, 45–48. See also Lehman, *Victorian Women and the Theatre of Trance*, 149–59.

4. Tylor, *Primitive Culture*, 1:384–86. Tylor was not the first scientific or medical practitioner to investigate the claims of spiritualists. For more see Walkowitz, "Science and the Séance," 3–29.

5. E. B. Tylor to A. R. Wallace, November 26, 1866, catalogue no. Add 46439 ff. 6, E. B. Tylor Papers, British Library, London.

6. E. B. Tylor to A. R. Wallace, November 26, 1866.

7. For more on collective empiricism see Daston and Galison, *Objectivity*, 19–27.

8. For more on the establishment of credible secondhand accounts in nineteenth-century anthropological and ethnographic studies see Sera-Shriar, "Tales from Patagonia," 209.

9. Stocking, "Animism in Theory and Practice," 92.

10. Alfred Russel Wallace described at length the spirit hypothesis in *On Miracles and Modern Spiritualism*, 100–117.

11. Tylor, *Primitive Culture*, 1:383.

12. The most noteworthy examples in the historiography are as follows: Leopold, *Culture in Comparative and Evolutionary Perspective*; Stocking, *Victorian Anthropology*, 299–319; Stocking, *After Tylor*; Strenski, *Thinking about Religion*, 91–116; Larsen, *Slain God*, 13–36; and Sera-Shriar, *Making of British Anthropology*, 147–76. There have been several articles and chapters published on aspects of Tylor's career and life including: Lang, "Edward Burnett Tylor," 1–17; Murphree, "Evolutionary Anthropologists," 265–300; Stocking, "'Cultural Darwinism,'" 91–109; Stringer, "Rethinking Animism," 541–55; Stocking, "Edward Burnett Tylor and the Mission of Primitive Man," 103–15; Barth, "Britain and the Commonwealth," 6–14, 22; Kuklick, "The British Tradition," 52–60; Ratnapalan, "E. B. Tylor and the Problem of Primitive Culture," 131–42; Sera-Shriar, "Historicizing Belief," 68–90.

13. Marett, *Tylor*.

14. Stocking, "Animism in Theory and Practice," 88–104. There is some passing reference to Tylor's engagement with spiritualism in the secondary literature. Two notable examples are: Oppenheim, *Other World*, 16, 61; and Luckhurst, *Invention of Telepathy*, 14, 27, 42, 160–61.

15. Bleichmar, *Visible Empire*, 6–10. For more on observational practices within Victorian anthropology see Sera-Shriar, "What Is Armchair Anthropology?" 180–94. For more secondary literature on the history of scientific observation within the natural and social sciences see Secord, "Artisan Naturalists," 135–206; Grasseni, *Skilled Visions*; Grimshaw, *Ethnographer's Eye*; and Daston and Galison, *Objectivity*.

16. I am borrowing the term "middling-sort" from E. P. Thompson, who convincingly argued that a recognizable "middle class," comparable to our contemporary understanding, was still developing during the eighteenth century. Thompson, "Patrician Society, Plebeian Culture," 382–405. A similar argument can be applied to middle-ranking society, of which Tylor's family was a part, during the first half of the nineteenth century.

17. Isichei, *Victorian Quakers*, 7; Cantor, *Quakers, Jews and Science*, 234–36.

18. Bravo, "Ethnological Encounters," 339–40. See also Laidlaw, "Heathens, Slaves and Aborigines," 134–61.

19. Barth, "Britain and the Commonwealth," 6–8.

20. Tylor, *Primitive Culture*, 1:383.

21. Tylor, *Primitive Culture*, 1:33. See also Ratnapalan, "E. B. Tylor and the Problem of Primitive Culture," 131–42.

22. Stocking, *Victorian Anthropology*, 157.

23. Tylor, *Anahuac*, 1.

24. James Cowles Prichard's version of monogenesis was the dominant form in British ethnological circles during the first half of the nineteenth century. For more see Sera-Shriar, "Human History and Deep Time," 19–22; and Sera-Shriar, *Making of British Anthropology*, 21–52.

25. Stocking, *Victorian Anthropology*, 157; and Sera-Shriar, *Making of British Anthropology*, 154–55.

26. Sera-Shriar, "Tales from Patagonia," 206–9.

27. Stocking, "What's in a Name?" 369–90; and Kenny, "From the Curse of Ham to the Curse of Nature," 363–88.

28. Stocking, *Victorian Anthropology*, 159.

29. For more on the anthropological schism of the 1860s see Burrow, "Evolution and Anthropology in the 1860's [*sic*]," 137–49; Stocking, "What's in a Name?" 369–90; Kenny, "From the Curse of Ham to the Curse of Nature," 363–88; Sera-Shriar, "Observing Human Difference," 461–91.

30. Kuklick, "British Tradition," 55–60.

31. Stocking, *Victorian Anthropology*, 191.

32. For more on Tylor's reading practices see Sera-Shriar, *Making of British Anthropology*, 164–65. See also Stocking, *Victorian Anthropology*, 156–64.

33. For more on Tylor's exit from the Society of Friends see Larsen, *Slain God*, 20.

34. Wheeler-Barclay, *Science of Religion in Britain*, 75.

35. Stocking, *Victorian Anthropology*, 190–91; and Sera-Shriar, "Historicizing Belief," 68–90. For a more nuanced understanding of the boundaries between science and religion in the nineteenth century see Turner, "Victorian Conflict between Science and Religion," 356–76; Brooke and Cantor, "Whose Science? Whose Religion?" 43–72; Lightman, "Victorian Sciences and Religion," 343–66; and Harrison, "'Science' and 'Religion,'" 81–106.

36. Stocking, "Animism in Theory and Practice," 89.

37. Tylor discusses the earlier versions of his theory of animism in the preface to *Primitive Culture*, 1:v.

38. Tylor discusses the links between spiritualism and primitive belief at length in his lecture at the Royal Institution in 1869. See Tylor, "On the Survival of Savage Thought," 522–35.

39. For more on miracles and its links to spiritualism see Lamont, *Extraordinary Beliefs*, 144, 162–64. See also Taves, *Fits, Trances, and Visions*.

40. For more on the exaggerated nature of spiritualist writings see Bann, "Ghostly Hands and Ghostly Agency," 666.

41. Tylor, "On the Survival of Savage Thought," 526.

42. Tylor, "On the Survival of Savage Thought," 526.

43. Tylor, "On the Survival of Savage Thought," 526.

44. Tylor, "On the Survival of Savage Thought," 526.

45. [Gloumeline], *D.D. Home*, 299–300. See also Lamont, *First Psychic*, 183–96.

46. Lamb, *Victorian Magic*, 60–61; Porche and Vaughan, *Psychics and Mediums in Canada*, 40; Byrne, *Modern Spiritualism and the Church of England*, 70; and Lamont, *Extraordinary Beliefs*, 140–41.

47. Tylor, "On the Survival of Savage Thought," 526. These inconsistencies in the narrative connect to Peter Lamont's concept of the "crisis of evidence" in Victorian spiritualism. When problems in the observations of witnesses emerged, it was easier for spiritualist explanations to gain priority. Supernaturalism did not require as much tangible evidence to support its plausibility. The onus of evidence was always on the skeptic. See Lamont, "Spiritualism and a Mid-Victorian Crisis of Evidence," 897–920.

48. Tylor, "On the Survival of Savage Thought," 526.

49. Tylor, "On the Survival of Savage Thought," 526.

50. Tylor, "On the Survival of Savage Thought," 526.

51. Tylor, "On the Survival of Savage Thought," 528.

52. For a general discussion on scientific naturalism in Victorian Britain see Dawson and Lightman, *Victorian Scientific Naturalism*. For more on the vocational strategies of scientific naturalists such as Thomas Henry Huxley see Caudill, "Bishop-Eaters," 441–60; White, *Thomas Huxley*, 51–58; and Elwick, *Styles of Reasoning in the British Life Sciences*, 131–59; Lightman, *Victorian Popularizers of Science*, 359–61; and Kaalund, "Oxford Serialized," 429–53.

53. Tylor, *Primitive Culture*, 1:2.

54. For more on the importance of belief in Tylor's theory of animism see Vasconcelos, "Homeless Spirits," 23–24.

55. Manias, "Problematic Construction of 'Paleolithic Man,'" 32–43. For more on the comparative method in nineteenth-century British sciences see Burrow, *Evolution and Society*; Leopold, *Culture in Comparative and Evolutionary Perspective*; and Bowler, *Invention of Progress*.

56. Tylor, *Primitive Culture*, 1:20–21.

57. Tylor, *Primitive Culture*, 1:383.

58. Wallace, *On Miracles and Modern Spiritualism*, 102.

59. Tylor, *Primitive Culture*, 1:387.

60. Wallace, *On Miracles and Modern Spiritualism*, 102; Tylor, *Primitive Culture*, 1:387.

61. Tylor, *Primitive Culture*, 1:408. See also Bann, "Ghostly Hands and Ghostly Agency," 667–69.

62. Wallace, *On Miracles and Modern Spiritualism*, 47. Wallace also wrote a particularly scathing review of *Primitive Culture* in 1872, in which he argued that Tylor's work was theoretically unsophisticated and mainly lumped data together without extrapolating sufficiently on its significance. See Wallace, review of *Primitive Culture*, 69–71.

63. Geertz, *Interpretation of Cultures*, 5–6.

64. For more on the politics of writing in anthropology see Clifford and Marcus, *Writing Culture*; and Vargas-Cetina, *Anthropology and the Politics of Representation*.

65. Bleichmar, *Visible Empire*, 6–10.

66. For more on Tylor's impact on the formation of Victorian anthropology see Stocking, *After Tylor*.

67. Crookes, *Researches in the Phenomena of Spiritualism*, 9.

68. Much of these investigations were later published in Crookes, *Researches in the Phenomena of Spiritualism*. For more on Crookes's experiments see Oppenheim, *Other World*, 338–54; Luckhurst, *Invention of Telepathy*, 24–36; Grimes, *Late Victorian Gothic*, 126–27; and chapter 8 of Brock, *William Crookes (1832–1919)*.

69. For more on the ways scientific practitioners used quantifiable evidence in their research to support and strengthen their theories see Gooday, *Morals of Measurement*. For more on empiricism in spirit investigations see Holloway, "Enchanted Spaces," 183.

70. For more on Crookes's credentials as an expert observer of physical phenomena, and its links to spiritualism see Noakes, "Haunted Thoughts of the Careful Experimentalist," 46–56 and *Physics and Psychics*, 206–9.

71. Stocking, "Animism in Theory and Practice," 92.

72. Stocking, "Animism in Theory and Practice," 92.

73. Lamont, "Spiritualism and a Mid-Victorian Crisis of Evidence," 899.

74. Stocking, "Animism in Theory and Practice," 92.

75. Stocking, "Animism in Theory and Practice," 93. See also Crookes, *Researches in the Phenomena of Spiritualism*, 68–71; and Ferguson, *Determined Spirits*, 11.

76. Bann, "Ghostly Hands and Ghostly Agency," 666.

77. Owen, *Darkened Room*, 66, 72.

78. For more on gender, class, and power dynamics in séance investigations see Walkowitz, "Science and the Séance," 3–29.

79. Stocking, "Animism in Theory and Practice," 102.

80. Stocking, "Animism in Theory and Practice," 93.

81. Stocking, "Animism in Theory and Practice," 93.

82. Stocking, "Animism in Theory and Practice," 93. For more on common practices in séances see Holloway, "Enchanted Spaces," 184.

83. Stocking, "Animism in Theory and Practice," 93.

84. Lamont, "Magician as Conjuror," 26.

85. Stocking, "Animism in Theory and Practice," 93.

86. Stocking, "Animism in Theory and Practice," 93.

87. Stocking, "Animism in Theory and Practice," 93.

88. Stocking, "Animism in Theory and Practice," 94.

89. Stocking, "Animism in Theory and Practice," 94.

90. Stocking, "Animism in Theory and Practice," 94.

91. For more on Cox's spirit investigations see Cox, *Spiritualism Answered by Science*. See also Hall, *Spiritualists*, 79–84.

92. Podmore, *Modern Spiritualism*, 2:127.

93. Owen, *Darkened Room*, 115–17. See also Hoare, *England's Lost Eden*, 350–51; Richardson, *Second Sight in the Nineteenth Century*, 103–49.

94. Stocking, "Animism in Theory and Practice," 94.

95. For more on spiritualism in the United States see Nartonis, "Rise of Nineteenth-Century American Spiritualism," 361–73.

96. Stocking, "Animism in Theory and Practice," 94. See also Luckhurst, "Occult Gazetteer of Bloomsbury," 53–54.

97. Stocking, "Animism in Theory and Practice," 94–95.

98. Stocking, "Animism in Theory and Practice," 95.

99. Stocking, "Animism in Theory and Practice," 95.

100. For more on hysteria and mediumship see Owen, *Darkened Room*, 143–50. More broadly, see Chodoff, "Hysteria and Women," 545–51.

101. Richardson, *Second Sight in the Nineteenth Century*, 110–14.

102. Nelson, *Spiritualism and Society*, 99–100; and Holloway, "Enchanted Spaces," 183.

103. "Mrs. Jennie Holmes's Séances," 436; and "Mrs. Olive, Trance Medium," 436.

104. Buckland, *Spirit Book*, 148.

105. Stocking, "Animism in Theory and Practice," 95.

106. Stocking, "Animism in Theory and Practice," 96.

107. Howitt, *History of the Supernatural in All Ages and Nations*.

108. Moses, *Direct Spirit Writing (Psychography)*, 109–12. See also Meade, *Madame Blavatsky*, 178–81; and Murphet, *When Daylight Comes*, 195.

109. Stocking, "Animism in Theory and Practice," 96.

110. Stocking, "Animism in Theory and Practice," 97. See also Whittington-Egan, *Mrs. Guppy Takes a Flight*, 197.

111. Stocking, "Animism in Theory and Practice," 97.

112. Stocking, "Animism in Theory and Practice," 97.

113. Stocking, "Animism in Theory and Practice," 97.

114. Stocking, "Animism in Theory and Practice," 97.

115. Walkowitz, "Science and the Séance," 5.

116. Clifford and Marcus, *Writing Culture*; and Vargas-Cetina, *Anthropology and the Politics of Representation*. See also Bleichmar, *Visible Empire*, 6–10.

117. Stocking, "Animism in Theory and Practice," 97–98.

118. For more on Fox's career see Davenport, *Death-Blow to Spiritualism*; Oppenheim, *Other World*, 11, 32, 35, 126, 297, 331, 344; Weisberg, *Talking to the Dead*; and Chapin, *Exploring Other Worlds*.

119. Stocking, "Animism in Theory and Practice," 98.

120. Stocking, "Animism in Theory and Practice," 98.

121. Stocking, "Animism in Theory and Practice," 98.

122. Stocking, "Animism in Theory and Practice," 98.

123. Stocking, "Animism in Theory and Practice," 98.

124. Stocking, "Animism in Theory and Practice," 99.

125. Oppenheim, *Other World*, 54–57, 65, 77–80; Owen, *Darkened Room*, 123, 197.

126. Stocking, "Animism in Theory and Practice," 99.

127. Stocking, "Animism in Theory and Practice," 99.

128. Moses, *Spirit Teachings*, vi, 4, 7.

129. Stocking, "Animism in Theory and Practice," 99.

130. Stocking, "Animism in Theory and Practice," 99.

131. Stocking, "Animism in Theory and Practice," 100.

132. Luckhurst, *Invention of Telepathy*, 72, 93, 100–102.

133. Elliotson, *Numerous Cases of Surgical Operation without Pain*; Elliotson, "Report of Various Trials of the Clairvoyance," 477–529; Elliotson, "Instances of Double States of Consciousness," 158–87.

134. Carpenter, "Electro-Biology and Mesmerism," 501–57; and Carpenter, *Mesmerism, Spiritualism, &c.*

135. Alvarado, "Dissociation in Britain during the Late Nineteenth Century," 9–33; and Cardeña and Winkelman, *Altering Consciousness*.

136. Stocking, "Animism in Theory and Practice," 100.

137. Stocking, "Animism in Theory and Practice," 100.

138. Tylor, *Primitive Culture*, 1:383.

139. For example, see Stocking, "Animism in Theory and Practice," 97, 99.

140. See Walkowitz, "Science and the Séance." 5.

Chapter 3: Andrew Lang

1. For more on Andrew Lang and his polymathic pursuits see Richardson, *Second Sight in the Nineteenth Century*, 151–82.

2. Quoted from Luckhurst, *Invention of Telepathy*, 161. The original quote comes from Reinach, "Andrew Lang," 309–19.

3. Lang, *Myth, Ritual and Religion*. For more on Tylorian animism see Murphree, "Evolutionary Anthropologists," 281–86; Stringer, "Rethinking Animism," 541–55; Strenski, *Thinking about Religion*, 93; and Ratnapalan, "E. B. Tylor and the Problem of Primitive Culture," 131–42.

4. James Moore was one of the first historians of science to apply the theory of cognitive dissonance to Victorian scientific figures, in *Post-Darwinian Controversies*, 14, 111–13, and 121. For more on cognitive dissonance see Festinger, *Theory of Cognitive Dissonance*; Stocking, *After Tylor*, 57–59; and Oppenheim, *Other World*, 1–3. See also Darwin, *On the Origin of Species*. For

more on the related concept of Victorian crisis of faith see Helmstadter and Lightman, *Victorian Faith in Crisis*; Larsen, *Crisis of Doubt*; Lane, *Age of Doubt*; and Franklin, *Spirit Matters*.

5. Luckhurst, *Invention of Telepathy*, 161.

6. Luckhurst, "Knowledge, Belief and the Supernatural at the Imperial Margin," 206–12.

7. Lang, *Cock Lane and Common Sense*, ix–x.

8. Most discussions of Lang's contributions to anthropology are rather short, and he is mainly positioned as an amateur who began as a follower of Tylor and later became a critic. See Stocking, *Victorian Anthropology*, 260–63, 287, 320; Stocking, *After Tylor*, 49–64, 81–82, 174–76, 310–13; Kuklick, *Savage Within*, 79, 88, 253; Sebastian Lecourt, *Cultivating Belief*, 163–95. The most sustained discussions of his work in psychical research appear in Luckhurst, *Invention of Telepathy*, 160–67; Richardson, *Second Sight in the Nineteenth Century*, 151–82; and Raia, *New Prometheans*, 265–303. There is some scattered mention of Lang in Davies, *Haunted*, 8, 165; Vasconcelos, "Homeless Spirits," 25–27; and McCorristine, *Spectres of the Self*, 206–9.

9. Green, *Andrew Lang*.

10. Lang, "Edward Burnett Tylor," 1.

11. For more on Lang's challenge to anthropology see: Stocking, *After Tylor*, 57–59; and Wheeler-Barclay, *Science of Religion in Britain*, 140.

12. Ginzburg, "Microhistory," 10–35; Levi, "On Microhistory," 93–113; Lepore, "Historians Who Love Too Much," 129–44; and Karl Appuhn, "Microhistory," *Encyclopedia of European Social History*, accessed March 16, 2020, https://www.encyclopedia.com/international/ encyclopedias-almanacs-transcripts-and-maps/microhistory.

13. Stocking, *After Tylor*, 50.

14. Stocking, *After Tylor*, 51.

15. Donaldson, "Lang, Andrew (1844–1912), Anthropologist, Classicist, and Historian," *Oxford Dictionary of National Biography*, accessed June 7, 2018, http://www.oxforddnb.com/ view/10.1093/ref:odnb/9780198614128.001.0001/odnb-9780198614128-e-34396.

16. McLennan, *Primitive Marriage*; and Tylor, *Primitive Culture*. For more on Tylorian anthropology, folklore, and myth see Murphree, "Evolutionary Anthropologists," 286.

17. In *Cock Lane and Common Sense*, Lang refers to the "anthropological method" and the "anthropological test of evidence" on pages ix, 131, 162, 318, and 355.

18. Stocking, *After Tylor*, 53–55; and Luckhurst, "Knowledge, Belief and the Supernatural," 206.

19. For more on Lang's critique of Müller see Luckhurst, *Invention of Telepathy*, 161; and Stocking, *After Tylor*, 55–56.

20. Lang, *Custom and Myth*; and Lang, *Myth, Ritual and Religion*. See also Stocking, *After Tylor*, 53–55. Lang's early exposure to Tylor's work occurred at Oxford in 1872. For more, see Crawford, "*Pater's Renaissance*, Andrew Lang, and Anthropological Romanticism," 849–79.

21. Bleichmar, *Visible Empire*, 6–10.

22. For more on observation in Victorian anthropology see Sera-Shriar, "What Is Armchair Anthropology?" 180–94. On collective empiricism see Daston and Galison, *Objectivity*, 19–27. See also Bleichmar's discussion on long-distance observation in *Visible Empire*, 66–72.

23. Stocking, *After Tylor*, 56. See also Murphree, "Evolutionary Anthropologists," 265–300; and Stocking, "Matthew Arnold, E. B. Tylor, and the Uses of Invention," 783–99.

24. For more on Lang and abnormal facts see Lang, *Cock Lane and Common Sense*, 338, 345.

25. Lang, "Comparative Study of Ghost Stories," 623–32. See also Luckhurst, *Invention of Telepathy*, 162.

26. Lang, "Comparative Study of Ghost Stories," 623–24.

27. Luckhurst, *Invention of Telepathy*, 161–62.

28. Lang, "Comparative Study of Ghost Stories," 624.

29. Lang, "Comparative Study of Ghost Stories," 624.

30. Lang, "Comparative Study of Ghost Stories," 624.

31. Lang, "Comparative Study of Ghost Stories," 624.

32. Lang, "Comparative Study of Ghost Stories," 624.

33. Lang, "Comparative Study of Ghost Stories," 624.

34. Lang, "Comparative Study of Ghost Stories," 625.

35. Lang, "Comparative Study of Ghost Stories," 632. Tylor discusses this material in Tylor, *Primitive Culture*, 1:377–453.

36. Lang, *Cock Lane and Common Sense*, ix.

37. Shapin and Schaffer, *Leviathan and the Air-Pump*, 55–65.

38. Lang, *Cock Lane and Common Sense*, xi.

39. The private letter was published posthumously in a biography written by Margaret Maria Gordon in 1869. For more on Brewster's investigation of D. D. Home see Gordon, *Home Life of Sir David Brewster*, 257–58; and [Gloumeline], *D. D. Home*, 36–44.

40. Lang, *Cock Lane and Common Sense*, xi.

41. Brewster, *Letters on Natural Magic*.

42. For more on Brewster and his *Letters on Natural Magic* see Morus, "Illuminating Illusions, or, the Victorian Art of Seeing Things," 37–50.

43. Lang, *Cock Lane and Common Sense*, xiii.

44. For more on Owen and spiritualism see Lehman, *Victorian Women and the Theatre of Trance*, 160–68. See also Owen, *Footfalls on the Boundary of Another World*; and Owen, *Debatable Land between This World and the Next*.

45. Lang, *Cock Lane and Common Sense*, xiv.

46. Lang, *Cock Lane and Common Sense*, 5.

47. Lang, *Cock Lane and Common Sense*, 5.

48. Lang, *Cock Lane and Common Sense*, 8.

49. Lang, *Cock Lane and Common Sense*, 10.

50. Thyraeus, *Loca Infesta, hoc est*.

51. Lang, *Cock Lane and Common Sense*, 131, 162, 318, and 355.

52. Lang, *Cock Lane and Common Sense*, 131–32.

53. For more on the Society for Psychical Research's investigations of ghosts see McCorristine, *Spectres of the Self*, 124–29.

54. Lang, *Cock Lane and Common Sense*, 136–37.

55. Lang, *Cock Lane and Common Sense*, 137.

56. Lang, *Cock Lane and Common Sense*, 137.

57. Lang, *Cock Lane and Common Sense*, 137–38.

58. Lang, *Cock Lane and Common Sense*, 138.

59. Lang, *Cock Lane and Common Sense*, 138.

60. Lang, *Cock Lane and Common Sense*, 139.

61. Lang, *Cock Lane and Common Sense*, 140.

62. Daston and Galison, *Objectivity*, 19–27.

63. Lang, *Cock Lane and Common Sense*, 144.

64. Lang, *Cock Lane and Common Sense*, 144.

65. Lang, *Cock Lane and Common Sense*, 144–45.

66. Lang, *Cock Lane and Common Sense*, 145.
67. For example, see Tylor, *Primitive Culture*, 1:408.
68. Gurney, Podmore, and Myers, *Phantasms of the Living*.
69. Lang, *Cock Lane and Common Sense*, 183.
70. Lang, *Cock Lane and Common Sense*, 184.
71. Lang, *Cock Lane and Common Sense*, 187.
72. Lang, *Cock Lane and Common Sense*, 145.
73. Lang, *Cock Lane and Common Sense*, 145.
74. Lang, *Cock Lane and Common Sense*, 145.
75. Lang, *Cock Lane and Common Sense*, 146.
76. For more on telepathic thought transference see chapter 4 of McCorristine, *Spectres of the Self*, 139–91. For more on the history of telepathy generally see Luckhurst, *Invention of Telepathy*.
77. Lang, *Cock Lane and Common Sense*, 146–47.
78. Lang, *Cock Lane and Common Sense*, 147.
79. Lang, *Cock Lane and Common Sense*, 149.
80. Myers published two important articles on telepathic thought transference, which Lang used in his analysis: Myers, "On Telepathic Hypnotism, and Its Relation to Other Forms of Hypnotic Suggestion," 127–88; and "On a Telepathic Explanation of Some So-Called Spiritualistic Phenomena," 217–37. See also Gurney and Myers, "Transferred Impressions and Telepathy," 437–52.
81. Lang, *Cock Lane and Common Sense*, 161.
82. Lang, *Cock Lane and Common Sense*, 161.
83. Lang, *Cock Lane and Common Sense*, 162.
84. "An Account of the Detection of the Imposture in Cock-Lane," 81; and Goldsmith, *Mystery Revealed*. Reprinted versions of the account from *Gentleman's Magazine* appeared in numerous other periodicals at the time.
85. Geertz, *Interpretation of Cultures*, 5–6.
86. Bleichmar, *Visible Empire*, 6–10.
87. For more on the Cock Lane ghost story see Grant, *Cock Lane Ghost*; and Chambers, *Cock Lane Ghost*.
88. Lang, *Cock Lane and Common Sense*, 163–64.
89. Lang, *Cock Lane and Common Sense*, 164.
90. Lang, *Cock Lane and Common Sense*, 164–65.
91. Lang, *Cock Lane and Common Sense*, 164–65. See also Grant, *Cock Lane Ghost*, 14–15; and Chambers, *Cock Lane Ghost*, 39–40.
92. Lang, *Cock Lane and Common Sense*, 165.
93. Chambers, *Cock Lane Ghost*, 80–87.
94. Grant, *Cock Lane Ghost*, 110–12.
95. Lang, *Cock Lane and Common Sense*, 166.
96. Lang, *Cock Lane and Common Sense*, 167. For more on Inuit mediumship see McCorristine, *Spectral Arctic*.
97. For Samuel Johnson's account of the investigation in February 1762 see Boswell and Malone, *Life of Samuel Johnson*, 220–21.
98. Lang, *Cock Lang and Common Sense*, 168–69. Newgate was a famous large prison in the city of London, located near the Old Bailey, the central criminal court.
99. Lang, *Cock Lang and Common Sense*, 169.

100. Lang, *Cock Lang and Common Sense*, 169.

101. Lang, *Cock Lang and Common Sense*, 170.

102. Lang, *Cock Lang and Common Sense*, 161.

103. Lang, *Cock Lang and Common Sense*, 170–71.

104. William Fletcher Barrett's account of the investigation was published in "Demons of Derrygonnelly," 692–705.

105. Lang, *Cock Lang and Common Sense*, 171.

106. Lang, *Cock Lang and Common Sense*, 172.

107. Lang, *Cock Lang and Common Sense*, 172.

108. The details of the trial can be found in Owen, *Footfalls on the Boundary of Another World*, 195–203, and Mirville, *Fragment d'un Ouvrage Inédit*. Lang also published another version of the trial in "Poltergeist at Cideville," 454–63.

109. Lang, *Cock Lane and Common Sense*, 275.

110. Lang, *Cock Lane and Common Sense*, 276.

111. Lang, *Cock Lane and Common Sense*, 276.

112. Lang, *Cock Lane and Common Sense*, 276–77.

113. Lang, *Cock Lane and Common Sense*, 277.

114. Lang, *Cock Lane and Common Sense*, 277.

115. Lang, *Cock Lane and Common Sense*, 278–79. Bishop was one of the women accused and convicted of witchcraft during the Salem witch trials between 1692 and 1693. For more on Bishop and the Salem witch trials see Hall, *Witch-Hunting in Seventeenth-Century New England*, 280–315; and Rosenthal, *Salem Story*.

116. Lang cites Mather, *Wonders of the Invisible World*, 131, 150.

117. For more examples of global comparisons of mediumship see Lang, *Cock Lane and Common Sense*, 40–42, 67, 75, 99–100.

118. A list of the main witnesses and excerpts from their court testimonies can be found in "Sorcellerie," 129–73.

119. Lang, *Cock Lane and Common Sense*, 279.

120. Lang, *Cock Lane and Common Sense*, 279–80.

121. For information on Gibotteau's experiments Lang consults Myers, "Subliminal Consciousness" (1893–1894), 28–31. Another version can be found in Podmore, *Apparitions and Thought-Transference*, 368–70.

122. Lang, *Cock Lane and Common Sense*, 280–82.

123. Lang, *Cock Lane and Common Sense*, 306.

124. Lang, *Cock Lane and Common Sense*, 307.

125. Lang, *Cock Lane and Common Sense*, 310.

126. Gasparin, *Science vs. Modern Spiritualism*.

127. Figuier, *Les Mystères de la Science*, 2:579.

128. Shapin and Schaffer, *Leviathan and the Air-Pump*, 60–65.

129. For more on unconscious muscular motion see Faraday, "Experimental Investigation of Table-Turning," 328–33; "Michael Faraday's Researches in Spiritualism," 145–50; Oppenheim, *Other World*, 327–28, 336–37; and Luckhurst, *Invention of Telepathy*, 26.

130. Lang, *Cock Lane and Common Sense*, 310.

131. Lang, *Cock Lane and Common Sense*, 311.

132. Lang, *Cock Lane and Common Sense*, 311.

133. Thury, *Les Tables Tournantes*.

134. Lang, *Cock Lane and Common Sense*, 311–12.

135. For more on Faraday's and Carpenter's respective studies on table turning see Faraday, "Experimental Investigation of Table-Turning," 328–33; "Michael Faraday's Researches in Spiritualism," 145–50; and Carpenter, "Electro-Biology and Mesmerism," 501–57. See also Oppenheim, *Other World*, 327–28, 336–37.

136. Lang, *Cock Lane and Common Sense*, 313. For more on trance see Taves, *Fits, Trances and Visions*.

137. Lang, *Cock Lane and Common Sense*, 314.

138. See Carpenter, "Spiritualism and Its Latest Converts," 301–53; and Carpenter, "Mesmerism, Odylism, Table-Turning and Spiritualism," 135–57. For more on Carpenter's experiments on table turning see Lamont, "Spiritualism and a Mid-Victorian Crisis of Evidence," 901–3.

139. Lang, *Cock Lane and Common Sense*, 319–20.

140. Lang, *Cock Lane and Common Sense*, 320. Wallace's critique of Carpenter appears in Carpenter, *Mesmerism, Spiritualism, &c. Historically & Scientifically Considered*, vii. According to Carpenter the debate arose during Wallace's chairmanship of the Anthropological Department at the 1876 meeting of the British Association for the Advancement of Science in Glasgow. For more see Slotten, *Heretic in Darwin's Court*, 326–37.

141. Lang, *Cock Lane and Common Sense*, 321.

142. Clodd, "Presidential Address," 79.

143. Clodd, "Presidential Address," 80–81.

144. See Lang, "Protest of a Psycho-Folklorist," 236–48; and Clodd, "Reply to the Foregoing 'Protest,'" 248–58.

Chapter 4: Edward Clodd

1. For more on Clodd's popular science writing see Lightman, "Darwin and the Popularization of Evolution," 5–24; and Lightman, *Victorian Popularizers of Science*, 253–66; Lorimer, "Science and the Secularization of Victorian Images of Race," 222–24; and Lorimer, *Science, Race Relations and Resistance*, 135–40.

2. For Clodd's remarks on the influence on Tylor's writings on his work see Clodd, *Memories*, 16. See also Clodd, *Childhood of the World*; and Clodd, *Childhood of Religions*.

3. One of the most famous accounts of a grieving family trying to communicate with a deceased loved one is Lodge, *Raymond, or, Life and Death*. For more on spiritualism during and after the First World War see Hazelgrove, "Spiritualism after the Great War," 404–30; Hazelgrove, *Spiritualism and British Society between the Wars*; and Davies, *Supernatural War*.

4. Clodd, *Question*.

5. Clodd delivered these two lectures on May 17 and 24, 1921, and they were published the following year as Clodd, *Occultism*.

6. Clodd, "Reply to the Foregoing 'Protest,'" 248.

7. Clodd, *Question*, preface (n.p.).

8. Daston and Galison, *Objectivity*, 39–42.

9. Lightman, *Victorian Popularizers of Science*, 253–66.

10. Lorimer, *Science, Race Relations and Resistance*, 135–40.

11. For example, there is no mention of Clodd in any of the following sources: Oppenheim, *Other World*; Lamont, *Extraordinary Beliefs*; and McCorristine, *Spectres of the Self*. There is sparse reference to Clodd in Luckhurst, *Invention of Telepathy*, 163, 196, and 198; and Richardson, *Second Sight in the Nineteenth Century*, 159, 162, and 185.

12. McCabe, *Edward Clodd*, v.

13. McCabe, *Edward Clodd*, 31–32.

14. For more on the institutionalization of anthropology in the early twentieth century see Kuklick, *Savage Within*, 27–28; Stocking, *Ethnographer's Magic and Other Essays*, 12–59; Kuklick, "After Ishmael," 47; and Sera-Shriar, *Making of British Anthropology*, 2–3, 178–79.

15. For more on scientific naturalism and agnosticism see Lightman, *Origins of Agnosticism* and "Huxley and Scientific Agnosticism," 271–89.

16. Clodd, *Memories*, 2.

17. Clodd, *Memories*, 3–5.

18. Clodd, *Memories*, 7–8.

19. Clodd, *Memories*, 8.

20. Clodd, *Memories*, 8–9.

21. Darwin, *On the Origin of Species*; and Kirchhoff and Bunsen, "Chemical Analysis by Observation of Spectra," 161–89.

22. Clodd, *Memories*, 12.

23. Much has been written on the Oxford debate of 1860, but the most detailed account is Hesketh, *Of Apes and Ancestors*. See also Caudill, "Bishop-Eaters," 441–60; and Kaalund, "Oxford Serialized," 429–53.

24. Clodd, *Memories*, 16. See also Jowett, "On the Interpretation of Scripture,", 330–433. For more on Jowett's writings see Barr, "Jowett and the 'Original Meaning' of Scripture," 433–37.

25. Clodd, *Memories*, 14. See also [Seeley], *Ecce Homo*. For more on the anonymous publication of Seeley's *Ecce Homo*, and its impact on Victorian Britain see Hesketh, *Victorian Jesus*.

26. Huxley, *Evidence as to Man's Place in Nature*; and Huxley, *On Our Knowledge of the Causes of Organic Nature*. For more on Huxley's workingmen lectures see Lightman, *Victorian Popularizers of Science*, 353–54.

27. Clodd, *Memories*, 16–17.

28. Clodd, *Memories*, 17. See also Tylor, *Primitive Culture*. For more on Tylor and the evolution of religion see Sera-Shriar, "Historicizing Belief," 68–90. For more on the impact of Tylor's work on the discipline of anthropology generally see Stocking, *After Tylor*.

29. Lorimer, *Science, Race Relations and Resistance*, 136.

30. For more on Clodd as a popularizer see Lightman, *Victorian Popularizers of Science*, 253–66.

31. Clodd, *Memories*, 17.

32. Clodd, *Childhood of the World*; and Clodd, *Childhood of Religions*. See also Lightman, *Victorian Popularizers of Science*, 56–57; and Lorimer, *Science, Race Relations and Resistance*, 137–38.

33. Lorimer, *Science, Race Relations and Resistance*, 136. For more on scientific materialism see Gregory, *Scientific Materialism in Nineteenth Century Germany*.

34. Clodd, "Presidential Address," 78–79.

35. Clodd, "Presidential Address," 79.

36. Clodd, "Presidential Address," 80–81.

37. Clodd, "Presidential Address," 80.

38. Lang, "Protest of a Psycho-Folklorist," 236.

39. Lang, "Protest of a Psycho-Folklorist," 236.

40. Lang, "Protest of a Psycho-Folklorist," 236.

41. Lang, "Protest of a Psycho-Folklorist," 237.

42. Lang, "Protest of a Psycho-Folklorist," 237.

43. Lang, "Protest of a Psycho-Folklorist," 238.

44. For more on ideomotor responses and table turning see Faraday, "Experimental Investigation of Table-Turning," 328–33; "Michael Faraday's Researches in Spiritualism," 145–50; and Carpenter, "Electro-Biology and Mesmerism," 501–57. See also Oppenheim, *Other World*, 327–28, 336–37.

45. Lang, "Protest of a Psycho-Folklorist," 238–39.

46. Lang, "Protest of a Psycho-Folklorist," 241.

47. Clodd, "Reply to the Foregoing 'Protest,'" 248.

48. Clodd, "Reply to the Foregoing 'Protest,'" 248.

49. Clodd, "Reply to the Foregoing 'Protest,'" 248.

50. Clodd, "Reply to the Foregoing 'Protest,'" 249.

51. Clodd, "Reply to the Foregoing 'Protest,'" 250.

52. Clodd, "Reply to the Foregoing 'Protest,'" 250. The SPR's report on hallucination was published as Sidgwick et al., "Report on the Census of Hallucinations," 25–422. Most of the data that Clodd uses can be found in a table on page 36 of the document.

53. Clodd, "Reply to the Foregoing 'Protest,'" 249.

54. For more on Stead's spiritualism and editorship of *Borderland* see Oppenheim, *Other World*, 33–34; and Tuckett, *Evidence for the Supernatural*, 52–53. For more on Freer see Hall, *Strange Case of Ada Goodrich-Freer*; Luckhurst, *Invention of Telepathy*, 130; Campbell and Hall, *Strange Things*; and Richardson, *Second Sight in the Nineteenth Century*, 195–244.

55. Clodd, "Reply to the Foregoing 'Protest,'" 250.

56. Creese, *Ladies in the Laboratory?*, 109–10.

57. Clodd, "Reply to the Foregoing 'Protest,'" 250.

58. Clodd, "Reply to the Foregoing 'Protest,'" 253.

59. Bleichmar, *Visible Empire*, 6–10.

60. I am borrowing the term "skilled vision" from Cristina Grasseni, who uses it to mean a nuanced form of observation that is not simply the physical act of looking at an object of study but something more rigorous involving multiple senses, expert practices, and specialized theories. For more see Grasseni, "Introduction," 1–19.

61. Clodd, *Question*, 13.

62. For more on personal equations in the human sciences see Schaffer, "Astronomers Mark Time," 115–45; and Kuklick, "Personal Equations," 1–33.

63. For more on collective empiricism see Daston and Galison *Objectivity*, 19–27.

64. Clodd, *Question*, 14–15. Frederic W. H. Myers referred to the disassociated part of the mind as the "subliminal self." For more see Myers, "Subliminal Consciousness," 298–355; Oppenheim, *Other World*, 257–62; Finn, *Figures of the Pre-Freudian Unconscious*, 17–18.

65. Clodd, *Question*, 15.

66. Clodd, *Question*, 15.

67. Clodd, *Question*, 29. Clodd was quoting Murray, *Four Stages of Greek Religion*, 111.

68. For more on the burned-over district see Cross, *Burned-over District*; Pritchard, "Burned-Over District Reconsidered," 243–65; and Wellman, *Grassroots Reform in the Burned-Over District*.

69. Clodd, *Question*, 33–39.

70. Clodd, *Question*, 68. See also Goldsmith, *Mystery Revealed*; and "Account of the Detection of the Imposture in Cock-Lane," 81.

71. Clodd, *Question*, 69. Original source: Wright, *Letters of Horace Walpole*, 3:479.

72. Clodd, *Question*, 69–70.

73. Clodd, *Question*, 70. Original source: Wright, *Letters of Horace Walpole*, 3:381–82.

74. Clodd, *Question*, 71–72.

75. Clodd, *Question*, 71–72.

76. Clodd, *Question*, 72. Clodd's source for Johnson's account was "Account of the Detection of the Imposture in Cock-Lane," 81.

77. Clodd, *Question*, 73–74.

78. Lamont, "Spiritualism and a Mid-Victorian Crisis of Evidence," 899.

79. Podmore, *Modern Spiritualism*, 2:14.

80. Stocking, "Animism in Theory and Practice," 92.

81. Clodd, *Question*, 42.

82. Clodd, *Question*, 42. Clodd stated that Home received in excess of £24,000 from Lyons, while historian Amy Lehman has argued that it was closer to £60,000. See Lehman, *Victorian Women and the Theatre of Trance*, 145.

83. Home's second wife wrote a biography about his life: [Gloumeline], *D. D. Home*.

84. Clodd, *Question*, 87.

85. Clodd, *Question*, 43.

86. [Gloumeline], *D. D. Home*, 299–300, and Lamont, *First Psychic*, 183–96.

87. Clodd, *Question*, 46.

88. Crookes discussed his notion of an unknown force in *Researches in the Phenomena of Spiritualism*, 9. See also Oppenheim, *Other World*, 345.

89. Clodd, *Question*, 88.

90. Clodd, *Question*, 88.

91. Clodd, *Question*, 96.

92. Clodd, *Question*, 96.

93. Clodd, *Question*, 112.

94. Wallace, *On Miracles and Modern Spiritualism*, 102; and Lang, *Cock Lane and Common Sense*, 79–80.

95. Clodd, *Question*, 53.

96. Clodd, *Question*, 114.

97. Clodd, *Question*, 52–53. Original source: Wallace, *On Miracles and Modern Spiritualism*, 102.

98. Clodd, *Question*, 111.

99. Oliver Lodge also believed that Moses had genuinely sensed the death of the Baker Street suicide victim. See Lodge, *Raymond, or, Life and Death*, 350.

100. Clodd, *Question*, 112.

101. Clodd, *Question*, 112.

102. For more on the Davenport brothers and the spirit cabinet trick, see Nichols, *Biography of the Brothers Davenport*, 55–61; and Houdini, *Magician among the Spirits*, 17–35; Lamont, *First Psychic*, 149–67; and Lamont, *Extraordinary Beliefs*, 131–41, 158–62.

103. Lamont, "Magician as Conjuror," 21–33.

104. Clodd, *Question*, 48–49.

105. Nichols, *Biography of the Brothers Davenport*, 206.

106. Nichols, *Biography of the Brothers Davenport*, 208–9.

107. Clodd, *Question*, 101–2.

108. "Davenports at Liverpool," 327. For more on the Davenport brothers' ill-fated trip to Britain between 1864 and 1865 see *Davenport Brothers, The World-Renowned Spiritual Mediums*, 395–410.

109. Cooper, *Spiritual Experiences*.

110. Clodd, *Question*, 102.

111. Podmore, *Modern Spiritualism*, 2:61.

112. Collingwood, "Anthropological Society and the Davenports," 393–96.

113. Coleman was best known for his book *Spiritualism in America*.

114. Clodd, *Question*, 102–3. Maskelyne had already reproduced the Davenports' act as early as 1865 at Cheltenham, Gloucestershire, but it received much broader attention at the Crystal Palace. For more on Maskelyne's performance see Podmore, *Modern Spiritualism*, 2:61; and Steinmeyer, *Hiding the Elephant*, 95–96, 201.

115. Clodd, *Question*, 50.

116. Truesdell, *Bottom Facts Concerning the Science of Spiritualism*, 143–61.

117. Evans, *Hours with the Ghosts*, 36; and Kurtz, *Skeptic's Handbook of Parapsychology*, 253.

118. For more on Slade's exposure by Lankester and Donkin see McCabe, *Spiritualism*, 160–61.

119. E. Ray Lankester and Horatio B. Donkin, "A Spirit-Medium," *Times*, September 16, 1876, 7; Lankester and Donkin, "A Spirit Medium," *Times*, September 21, 1876, 3; Lankester and Donkin, "A Sprit Medium," *Times*, September 23, 1876, 9.

120. Lester, *E. Ray Lankester and the Making of Modern Biology*, 93–104.

121. For more on Florence Cook see Owen, *Darkened Room*, 45–48; Oppenheim, *Other World*, 17–21; and Lehman, *Victorian Women and the Theatre of Trance*, 149–59.

122. Clodd, *Question*, 126.

123. Owen, *Darkened Room*, 66–69.

124. Clodd, *Question*, 127.

125. For more on nineteenth-century spirit photography and issues of credibility and trust see Tucker, *Nature Exposed*, 84–120.

126. Truesdell, *Bottom Facts Concerning the Science of Spiritualism*, 138–40.

127. Clodd, *Question*, 132

128. For more on Mumler and spirit photography see Kaplan, *Strange Case of William Mumler*; Carrington, *Personal Experiences in Spiritualism*, 208–9; Natale, *Supernatural Entertainments*, 134–69; and Hamer, "Helen F. Stuart and Hannah Frances Green," 146–67.

129. Clodd, *Occultism*.

130. For more on these disciplinary shifts in British anthropology during the early twentieth century see Sera-Shriar, *Making of British Anthropology*, 1–20 and 177–89. See also Urry, "Notes and Queries on Anthropology," 45–57; Urry, "History of Field Methods," 27–61; Tedlock, "From Participant Observation to the Observation of Participation," 61–94; Stocking, *Ethnographer's Magic and Other Essays*; Gupta and Ferguson, "Discipline and Practice," 1–46; Kuklick, "After Ishmael," 47–65; and Kuklick, "Personal Equations," 1–33. For more on Malinowski's career see Kuper, *Anthropology and Anthropologists*, 1–35.

131. Much of the rhetoric in the early twentieth century that sought to discredit Victorian anthropology was part of the professionalizing strategies of British academic anthropologists, who were attempting to forge new departments in the expanding university sector. As a way of making a case for themselves as specialist practitioners of a sophisticated science, they distanced themselves from the methods and theories of earlier researchers, whom they framed as amateur enthusiasts. For more on this period of professional and institutional transition in anthropology see Sera-Shriar, *Making of British Anthropology*, 1–20; Urry, "History of Field Methods," 27–61; Kuklick, "After Ishmael," 47–65; and Kuklick, "Personal Equations," 1–33.

132. Hazelgrove, "Spiritualism after the Great War," 404–30; Hazelgrove, *Spiritualism and British Society between the Wars*; and Davies, *Supernatural War*.

133. Lang, "Protest of a Psycho-Folklorist," 236.

134. The two obvious examples of skeptics engaging with Clodd's antispiritualist work are the magicians Harry Houdini and Joseph Francis Rinn, in Rinn, *Sixty Years of Psychical Research*, 175–80.

Epilogue: Legacies of Late Victorian Spirit Investigations

1. The most detailed studies of Houdini's engagement with spiritualism in the secondary literature are Sandford, *Houdini and Conan Doyle* and Noyes, *Magician and the Spirits*. A tremendous amount has been written on Harry Houdini and his career as a stage magician. Recent biographies include Kalush and Sloman, *Secret Life of Houdini*; Mullin, *Harry Houdini*; Savory, *Amazing Harry Houdini*; Weaver, *Harry Houdini*.

2. Houdini, *Magician among the Spirits*, xi.

3. Houdini, *Magician among the Spirits*, xi.

4. Houdini, *Magician among the Spirits*, xi.

5. Houdini, *Magician among the Spirits*, xii.

6. Houdini, *Magician among the Spirits*, xii.

7. Bleichmar, *Visible Empire*, 6–10.

8. Lang, "Protest of a Psycho-Folklorist," 236.

9. For more on these disciplinary shifts in British anthropology during the early twentieth century see Sera-Shriar, *Making of British Anthropology*, 1–20 and 177–89. See also Urry, "Notes and Queries on Anthropology," 45–57; Urry, "History of Field Methods," 27–61; Tedlock, "From Participant Observation to the Observation of Participation," 61–94; Stocking, *Ethnographer's Magic and Other Essays*; Gupta and Ferguson, "Discipline and Practice," 1–46; Kuklick, "After Ishmael," 47–65; and Kuklick, "Personal Equations," 1–33. For more on Malinowski's career see Kuper, *Anthropology and Anthropologists*, 1–35.

10. Houdini, *Magician among the Spirits*, xx.

11. Although Houdini does not explicitly state his religion in his writings, he was born Jewish.

12. Houdini, *Magician among the Spirits*, xviii–xix. Between 1922 and 1924 Houdini also made two scrapbooks on spiritualism that included images, newspaper and periodical clippings, advertisements, and other paraphernalia. These collections provide a window into the investigatory practice of Houdini. See Harry Houdini, "Spiritualism and Magic," circa 1922–1924, Harry Houdini Papers, Harry Ransom Center, University of Texas at Austin https://hrc.contentdm .oclc.org/digital/collection/p15878coll22/id/3076/rec/9; and Harry Houdini, "Spiritualism Scrapbook," circa 1922–1924, Harry Houdini Papers, https://hrc.contentdm.oclc.org/digital/collection /p15878coll22/id/3760/rec/10.

13. Houdini, *Magician among the Spirits*, xvii.

14. Two obvious examples were Crookes, *Researches in the Phenomena of Spiritualism*; and Wallace, *On Miracles and Modern Spiritualism*.

15. Lodge, *Raymond, or, Life and Death*; and Doyle, *New Revelation*. For more on the impact of Lodge and Doyle on the spiritualist movement see Sword, *Ghostwriting Modernism*.

16. Houdini, *Magician among the Spirits*, xvii–xviii.

17. For more on spiritualism during and after the First World War see Hazelgrove, "Spiritualism after the Great War," 404–30; Hazelgrove, *Spiritualism and British Society between the Wars*; and Davies, *Supernatural War*.

18. Houdini, *Magician among the Spirits*, xvi.

19. Houdini, *Magician among the Spirits*, xiv.

20. I am once again borrowing the term from Daniela Bleichmar. However, much of this discussion links to the history of scientific observation. For more see Bleichmar, *Visible Empire*, 6–10. For more secondary literature on the history of scientific observation within the natural and social sciences see Secord, "Artisan Naturalists," 135–206; Grasseni, *Skilled Visions*; Grimshaw, *Ethnographer's Eye*; and Daston and Galison, *Objectivity*.

21. Houdini, *Magician among the Spirits*, xv.

22. The most noteworthy examples in the historiography are Stocking, "'Cultural Darwinism' and 'Philosophical Idealism,'" 91–109; Stocking, "Animism in Theory and Practice," 88–104; Leopold, *Culture in Comparative and Evolutionary Perspective*; Stringer, "Rethinking Animism," 541–55; Strenski, *Thinking about Religion*, 91–116; Ratnapalan, "E. B. Tylor and the Problem of Primitive Culture," 131–42; Sera-Shriar, *Making of British Anthropology*, 147–76; and Sera-Shriar "Historicizing Belief," 68–90.

23. For more on Malinowki's anthropological program see Kuper, *Anthropology and Anthropologists*, 1–35; and Kuklick, "Personal Equations," 1–33. For more Haddon's field-centric approach see Kuklick, "After Ishmael," 47; and Kuklick, "British Tradition," 63. See also Stocking, *Victorian Anthropology*, 262–63; and Sera-Shriar, *Making of British Anthropology*, 1–20.

24. Quiggin, *Haddon, the Head Hunter*, 77.

25. Kuklick, "After Ishmael," 47; and Sera-Shriar, *Making of British Anthropology*, 2.

26. See, for example, the following sources: Hallowell, "Ojibwa Ontology, Behavior, and World View," 17–49; Bird-David, "'Animism' Revisited," 67–91; Harvey, *Animism*.

27. For more on these disciplinary transitions see Urry, "History of Field Methods," 27–61; Tedlock, "From Participant Observation to the Observation of Participation," 61–94; Gupta and Ferguson, "Discipline and Practice," 1–47; and Sera-Shriar, *Making of British Anthropology*, 1–20

28. For more on armchair anthropology in the nineteenth century see Sera-Shriar, "What Is Armchair Anthropology?" 180–94.

29. Much has been written on the so-called Victorian crisis of faith. For examples of the larger historiographical picture see Larsen, *Crisis of Doubt*; Schlossberg, *Conflict and Crisis in the Religious Life of Late Victorian England*; Lane, *Age of Doubt*. For more examples of specific literature on the Victorian crisis of faith and nineteenth-century science see Turner, "Victorian Conflict between Science and Religion," 356–76; Lightman, *Origins of Agnosticism*; Helmstadter and Lightman, *Victorian Faith in Crisis*; Turner, *Contesting Cultural Authority*; Brooke and Cantor, "Whose Science? Whose Religion?" 43–72; Lightman, "Victorian Sciences and Religion," 343–66; Sera-Shriar, "Historicizing Belief," 68–90. For more on modern spiritualism and the Victorian crisis of faith see Oppenheim, *Other World*, 1–3; and Franklin, *Spirit Matters*.

30. For a more nuanced understanding of the boundaries between science and religion see Turner, "Victorian Conflict between Science and Religion," 356–76; Brooke and Cantor, "Whose Science? Whose Religion?" 43–72; Harrison, "'Science' and 'Religion,'" 81–106; Lightman, "Victorian Sciences and Religion," 343–66; and Lightman, "Does the History of Science and Religion Change," 149–68.

31. Hesketh, *Of Apes and Ancestors*, 8–9; and Lyons, *Species, Serpents, Spirits, and Skulls*, 1–3.

32. For more on rationalistic doubt in the Victorian age see Larsen, *Crisis of Doubt*; Lane, *Age of Doubt*; and Franklin, *Spirit Matters*. For more on cognitive dissonance see Moore, *Post-Darwinian Controversies*, 14, 111–13, and 121; Festinger, *Theory of Cognitive Dissonance*; Stocking, *After Tylor*, 57–59; and Oppenheim, *Other World*, 1–3.

33. Ginzburg, "Microhistory," 10–35; Levi, "On Microhistory," 93–113; and Lepore, "Historians Who Love Too Much," 129–44; and Geertz, *Interpretation of Cultures*, 3–33. For more on observational practices in anthropology see Sera-Shriar, *Making of British Anthropology*, 1–20.

34. Wallace, *On Miracles and Modern Spiritualism*, 48–49. See also Wallace, "Spiritualism," 9:645–49.

35. Wallace, *On Miracles and Modern Spiritualism*, 100–109.

36. Tylor discusses his commitment to "prima facie evidence" in his notebook on spiritualism. See Stocking, "Animism in Theory and Practice," 100.

37. Lang's support of telepathy as a possible explanation of supposed spirit phenomena throughout history and cultures is discussed in detail in Lang, *Cock Lane and Common Sense*.

38. Clodd's disbelief in spiritualism and telepathy form the subject of his *Question* and *Occultism*.

39. I am borrowing the term "visual epistemologies" from Bleichmar, *Visible Empire*, 6–10.

40. For more secondary literature on the history of scientific observation and visual culture within the natural and social sciences see Secord, "Artisan Naturalists," 135–206; Grasseni, *Skilled Visions*; Grimshaw, *Ethnographer's Eye*; Daston and Galison, *Objectivity*; Daston and Lunbeck, *Histories of Scientific Observation*; and Sera-Shriar, *Making of British Anthropology*.

41. Shapin and Schaffer, *Leviathan and the Air-Pump*, 55–65.

42. For examples of literature on types of personal testimony as evidence in science see Golan, *Laws of Men and Laws of Nature*; Browne, "Science of Empire," 453–75; Bravo, "Ethnological Encounters," 338–57; Koerner, "Purposes of Linnaean Travel," 117–52; Carey, "Compiling Nature's History," 269–92; Hulme and Youngs, "Introduction," 1–16; Sera-Shriar, "Arctic Observers," 23–31; Sera-Shriar, "Tales from Patagonia," 204–23; and Kaalund, "From Science in the Arctic to Arctic Science."

43. For more on the history of truth and accuracy see Berger and Luckmann, *Social Construction of Reality*; Shapin, *Social History of Truth*; Appleby, Hunt, and Jacob, *Telling the Truth about History*; Hacking, *Social Construction of What?*; Dooley, *Social History of Skepticism*; Shapin, *Scientific Life*; Shapin, *Never Pure*; Baggini, *Short History of Truth*.

44. Noakes, *Physics and Psychics*, 8–9. See also Gieryn, *Cultural Boundaries of Science*; and Visvanathan, "Alternative Science," 164–69.

BIBLIOGRAPHY

"An Account of the Detection of the Imposture in Cock-Lane." *Gentleman's Magazine* 32 (1762): 81.

Adas, Michael. *Machines as the Measure of Men: Science, Technology, and Ideologies of Western Dominance*. Ithaca, NY: Cornell University Press, 1989.

Alexander, Sarah C. *Victorian Literature and the Physics of the Imponderable*. London: Pickering & Chatto, 2015.

Alvarado, Carlos S. "Dissociation in Britain during the Late Nineteenth Century." *Journal of Trauma and Dissociation* 3, no. 2 (2008): 9–33.

Appleby, Joyce, Lynn Hunt, and Margaret Jacob. *Telling the Truth about History*. New York: W. W. Norton, 1994.

Chéroux, Clément, Andreas Fischer, Pierre Apraxine, Denis Canguilhem, and Sophie Schmit. *The Perfect Medium: Photography and the Occult*. New Haven, CT: Yale University Press, 2005.

Asprem, Egil. *The Problem of Disenchantment: Scientific Naturalism and Esoteric Discourse, 1900–1939*. Leiden: Brill, 2014.

Baggini, Julian. *A Short History of Truth: Consolations for a Post-Truth World*. London: Quercus Books, 2017.

Bann, Jennifer. "Ghostly Hands and Ghostly Agency: The Changing Figure of the Nineteenth-Century Specter." *Victorian Studies* 51, no. 4 (2009): 663–85.

Barrow, Logie. *Independent Spirits: Spiritualism and English Plebeians, 1850–1910*. London: Routledge and Kegan Paul, 1986.

Barth, Fredrik. "Britain and the Commonwealth." In *One Discipline, Four Ways: British, German, French, and American Anthropology*, edited by Fredrik Barth, Andre Gingrich, Robert Parkin, and Sydel Silverman, 3–57. Chicago: University of Chicago Press, 2005.

Barton, Ruth. "'Huxley, Lubbock, and Half a Dozen Others': Professionals and Gentlemen in the Formation of the X Club, 1851–1864." *Isis* 89, no. 3 (1998): 410–44.

Barr, James. "Jowett and the 'Original Meaning' of Scripture." *Religious Studies* 18, no. 4 (1982): 433–37.

Barrett, Harrison Delivan. *The Life of Mrs. Cora L. V. Richmond*. Chicago: Hack & Anderson, 1895.

Barrett, William Fletcher. "The Demons of Derrygonnelly." *Dublin University Magazine*, 90 (1877): 692–705.

Bennett, Edward. *The Society for Psychical Research: Its Rise and Progress and a Sketch of Its Work*. London: R. Brimley Johnson, 1903.

Bennett, Joshua. *God and Progress: Religion and History in British Intellectual Culture, 1845–1914*. Oxford: Oxford University Press, 2019.

Berger, Peter L., and Thomas Luckmann. *The Social Construction of Reality: A Treatise in the Sociology of Knowledge*. New York: Anchor Books, 1966.

Bird-David, Nurit. "'Animism' Revisited: Personhood, Environment, and Relational Epistemology." *Current Anthropology* 40 (1991): 67–91.

Bleichmar, Daniela. *Visible Empire: Botanical Expeditions and Visual Culture in the Hispanic
Enlightenment*. Chicago: University of Chicago Press, 2012.

Blum, Deborah. *Ghost Hunters: William James and the Search for Scientific Proof of Life after
Death*. London: Penguin Books, 2006.

Boswell, James, and Edmond Malone. *The Life of Samuel Johnson*. London: Henry Baldwin, 1791.

Bowler, Peter. *The Invention of Progress: The Victorians and the Past*. Oxford: Basil Blackwell, 1989.

Bowler, Peter. *Reconciling Science and Religion: The Debate in Early Twentieth-Century Britain*.
Chicago: University of Chicago Press, 2001.

Brandon, Ruth. *The Spiritualists: The Passion for the Occult in the Nineteenth and Twentieth Centuries*. London: Weidenfeld & Nicholson, 1983.

Braude, Ann. *Radical Spirits: Spiritualism and Women's Rights in Nineteenth-Century America*.
Bloomington: Indiana University Press, 1989.

Bravo, Michael. "Ethnological Encounters." In *Cultures of Natural History*, edited by Nicho-
las Jardine, James Secord, and Emma Spary, 338–57. Cambridge: Cambridge University
Press, 1996.

Brewster, David. *Letters on Natural Magic, Addressed to Walter Scott*. London: John Murray, 1832.

Brock, William H. *William Crookes (1832–1919) and the Commercialization of Science*. Abingdon, UK: Routledge, 2008.

Brooke, John Hedley. *Science and Religion: Some Historical Perspectives*. Cambridge: Cambridge University Press, 2014.

Brooke, John Hedley, and Geoffrey Cantor. "Whose Science? Whose Religion?" In *Reconstructing Nature: The Engagement of Science and Religion*, edited by John Hedley Brooke
and Geoffrey Cantor, 43–72. Oxford: Oxford University Press, 1998.

Brown, Frank Burch. "The Evolution of Darwin's Theism." *Journal of the History of Biology* 19,
no. 1 (1986): 1–45.

Brown, Richard D. "Microhistory and the Post-modern Challenge." *Journal of the Early Re-
public* 23, no. 1 (2003): 1–20.

Browne, Janet. "Biogeography and Empire." In *Cultures of Natural History*, edited by Nicho-
las Jardine, James Secord, and Emma Spary, 305–21. Cambridge: Cambridge University
Press, 1996.

Browne, Janet. "A Science of Empire: British Biogeography before Darwin." *Revue d'Histoire
des Sciences* 45 (1992): 453–75.

Buckland, Raymond. *The Spirit Book: The Encyclopedia of Clairvoyance, Channeling, and Spirit
Communication*. Detroit: Visible Ink, 2005.

Burrow, John W. "Evolution and Anthropology in the 1860's [*sic*]: The Anthropological Society of London, 1863–71." *Victorian Studies* 7, no. 2 (1963): 137–49.

Burrow, John W. *Evolution and Society: A Study in Victorian Social Theory*. Cambridge: Cambridge University Press, 1966.

Byrne, Georgina. *Modern Spiritualism and the Church of England, 1850–1939*. Woodbridge,
UK: Boydell, 2010.

Campbell, John Lorne, and Trevor H. Hall. *Strange Things: Father Allan, Ada Goodrich Freer,
and the Second Sight*. Glasgow: Birlinn, 2006.

Cannon, Susan Faye. *Science in Culture: The Early Victorian Period*. New York: Science History, 1978.

Cantor, Geoffrey. *Quakers, Jews and Science: Religious Responses to Modernity and the Sciences in Britain, 1650–1900*. Oxford: Oxford University Press, 2005.

Cardeña, Etzel, and Michael Winkelman, eds. *Altering Consciousness: Multidisciplinary Perspectives*. Santa Barbara, CA: Praeger, 2011.

Carey, Daniel. "Compiling Nature's History: Travellers and Travel Narratives in the Early Royal Society." *Annals of Science* 54 (1997): 269–92.

Carpenter, William Benjamin. "Electro-Biology and Mesmerism." *Quarterly Review* 93 (1853): 501–57.

Carpenter, William Benjamin. "Mesmerism, Odylism, Table-Turning and Spiritualism, Considered Historically and Scientifically." *Fraser's Magazine* 15 (1877): 135–57.

Carpenter, William Benjamin. *Mesmerism, Spiritualism, &c. Historically & Scientifically Considered, Being Two Lectures Delivered at the London Institution*. London: Longmans, Green, 1877.

Carpenter, William Benjamin. "Spiritualism and its Latest Converts." *Quarterly Review* 131 (1871): 301–53.

Carrington, Hereward. *Personal Experiences in Spiritualism: Including the Official Account and Record of American Palladino Séances*. London: T. Werner Laurie, 1913.

Caudill, Edward. "The Bishop-Eaters: The Publicity Campaign for Darwin and *On the Origin of Species*." *Journal of the History of Ideas* 55 (1994): 441–60.

Cerullo, John. *The Secularization of the Soul: Psychical Research in Modern Britain*. Philadelphia: Institute for the Study of Human Issues, 1982.

Chadwick, Owen. *The Victorian Church*. 2 vols. London: SCM, 1966.

Chambers, Paul. *The Cock Lane Ghost: Murder, Sex and Haunting in Dr Johnson's London*. Stroud, UK: Sutton, 2006.

[Chambers, Robert]. *The Vestiges of the Natural History of Creation*. London: John Churchill, 1844.

Chapin, David. *Exploring Other Worlds: Margaret Fox, Elisha Kent Kane, and the Antebellum Culture of Curiosity*. Amherst: University of Massachusetts Press, 2004.

Chodoff, Paul. "Hysteria and Women." *American Journal of Psychiatry* 139, no. 5 (1982): 545–51.

Clifford, James, and George E. Marcus, eds. *Writing Culture: The Poetics and Politics of Ethnography*. Berkeley: University of California Press, 1986.

Clodd, Edward. *The Childhood of Religions: Embracing a Simple Account of the Birth and Growth of Myths and Legends*. London: Henry S. King, 1875.

Clodd, Edward. *The Childhood of the World; A Simple Account of Man in Early Times*. London: Macmillan, 1873.

Clodd, Edward. *Memories*. London: Chapman and Hall, 1916.

Clodd, Edward. *Occultism*. London: Watts, 1922.

Clodd, Edward. "Presidential Address." *Folklore* 6 (1895): 54–82.

Clodd, Edward. *The Question: A Brief History and Examination of Modern Spiritualism*. London: Grant Richards, 1917.

Clodd, Edward. "A Reply to the Foregoing 'Protest.'" *Folklore* 6 (1895): 248–58.

Coleman, Benjamin. *Spiritualism in America*. London: F. Pitman, 1861.

Collingwood, J. F. "The Anthropological Society and the Davenports." *Human Nature: A Monthly Journal of Zoistic Science* 2 (August 1868): 393–96.

Collins, H. M., and T. J. Pinch. *Frames of Meaning: The Social Construction of Extraordinary Science*. London: Routledge, 1982.

Coon, Deborah. "Testing the Limits of Sense and Science: American Experimental Psychologists Combat Spiritualism, 1800–1920." *American Psychologist* 47 (1992): 143–51.

Cooper, Robert. *Spiritual Experiences, Including Seven Months with the Brothers Davenport.* London: Heywood, 1867.

Cottom, Daniel. *Abyss of Reason: Cultural Movements, Revelations, and Betrayals.* Oxford: Oxford University Press, 1991.

Cox, Edward William. *Spiritualism Answered by Science.* New York: H. L. Hinton, 1871.

Cox, Edward William. *What Am I? A Popular Introduction to the Study of Psychology.* 2 vols. London: Longman, 1873.

Crawford, Robert. "*Pater's Renaissance*, Andrew Lang, and Anthropological Romanticism." *English Literary History* 53 (1986): 849–79.

Creese, Mary R. S. *Ladies in the Laboratory? American and British Women in Science, 1800–1900: A Survey of Their Contributions to Research.* London: Scarecrow, 1998.

Crookes, William. *Researches in the Phenomena of Spiritualism.* London: J. Burn, 1874.

Cross, Whitney R. *The Burned-over District: The Social and Intellectual History of Enthusiastic Religion in Western New York, 1800–1850.* Ithaca, NY: Cornell University Press, 1950.

Darwin, Charles. *On the Origin of Species by Means of Natural Selection.* London: John Murray, 1859.

Daston, Lorraine, and Peter Galison. *Objectivity.* New York: Zone Books, 2007.

Daston, Lorraine, and Elizabeth Lunbeck, eds. *Histories of Scientific Observation.* Chicago: University of Chicago Press, 2011.

Davenport, Reuben Briggs. *The Death-Blow to Spiritualism: Being the True Story of the Fox Sisters, as Revealed by Authority of Margaret Fox Kane and Catherine Fox Jencken.* New York: G. W. Dillingham, 1888.

The Davenport Brothers, The World-Renowned Spiritual Mediums: Their Biography, and Adventures in Europe and America. Boston: William White, 1869.

"The Davenports at Liverpool." *Orchestra* 3 (1865): 327.

Davies, Owen. *The Haunted: A Social History of Ghosts.* Basingstoke: Palgrave Macmillan, 2007.

Davies, Owen. *A Supernatural War: Magic, Divination, and Faith during the First World War.* Oxford: Oxford University Press, 2019.

Davis, Natalie Zemon. *The Return of Martin Guerre.* Cambridge, MA: Harvard University Press, 1983.

Dawson, Gowan, and Bernard Lightman, eds. *Victorian Scientific Naturalism: Community, Identity, Continuity.* Chicago: University of Chicago Press, 2014.

Desmond, Adrian. "Redefining the X Axis: 'Professionals,' 'Amateurs' and the Making of Mid-Victorian Biology—A Progress Report," *Journal of the History of Biology* 34, no. 1 (2001): 3–50.

Desmond, Adrian, and James Moore. *Darwin's Sacred Cause: Race, Slavery and the Quest for Human Origins.* London: Penguin Books, 2009.

Dettelbach, Michael. "The Face of Nature: Precise Measurement, Mapping, and Sensibility in the World of Alexander von Humboldt." *Studies in History and Philosophy of Biological and Biomedical Sciences* 30 (1999): 473–504.

Diamond, Jared. *Guns, Germs, and Steel: The Fates of Human Societies.* London: W. W. Norton, 1997.

Dooley, Brendan. *The Social History of Skepticism: Experience and Doubt in Early Modern Culture.* Baltimore, MD: Johns Hopkins University Press, 1999.

Doyle, Arthur Conan. *The New Revelation*. London: Hodder & Stoughton, 1918.

Durant, John. "Scientific Naturalism and Social Reform in the Thought of Alfred Russel Wallace." *British Journal for the History of Science* 12 (1979): 31–58.

Elliotson, John. "Instances of Double States of Consciousness Independent of Mesmerism." *Zoist* 4 (1846): 158–87.

Elliotson, John. *Numerous Cases of Surgical Operation without Pain in the Mesmeric State*. Philadelphia: Lea and Blanchard, 1843.

Elliotson, John. "Report of Various Trials of the Clairvoyance of Alexis Didier Last Summer." *Zoist* 2 (1845): 477–529.

Elwick, James. *Styles of Reasoning in the British Life Sciences: Shared Assumptions, 1820–58*. London: Pickering & Chatto, 2007.

Endersby, Jim. *Imperial Nature: Joseph Hooker and the Practices of Victorian Science*. Chicago: University of Chicago Press, 2008.

Evans, Henry Ridgely. *Hours with the Ghosts, or, Nineteenth-Century Witchcraft*. Chicago: Laird and Lee, 1897.

Evans, Henry Ridgely. *The Spirit World Unmasked: Illustrated Investigations into the Phenomena of Spiritualism and Theosophy*. Chicago: Laird and Lee, 1897.

Faraday, Michael. "Experimental Investigation of Table-Turning." *Journal of the Franklin Institute* 56 (1853): 328–33.

Ferguson, Christine. *Determined Spirits: Eugenics, Heredity and Racial Regeneration in Anglo-American Spiritualist Writing, 1848–1930*. Edinburgh: Edinburgh University Press, 2012.

Ferguson, Christine. "Other Worlds: Alfred Russel Wallace and Cross-Cultural Spiritualism." *Victorian Review* 41, no. 2 (2015): 177–91.

Festinger, Leon. *A Theory of Cognitive Dissonance*. Stanford, CA: Stanford University Press, 1957.

Fichman, Martin. *An Elusive Victorian: The Evolution of Alfred Russel Wallace*. Chicago: University of Chicago Press, 2004.

Fichman, Martin. "Science in Theistic Contexts: A Case Study of Alfred Russel Wallace on Human Evolution." *Osiris* 16 (2001): 227–50.

Figuier, Louis. *Les Mystères de la Science*. 2 vols. Paris: Librairie Illustrée, 1880.

Finn, Michael R. *Figures of the Pre-Freudian Unconscious from Flaubert to Proust*. Cambridge: Cambridge University Press, 2017.

Flandreau, Marc. *Anthropologists in the Stock Exchange: A Financial History of Victorian Science*. Chicago: University of Chicago Press, 2016.

Flannery, Michael. "Alfred Russel Wallace, Nature's Prophet: From Natural Selection to Natural Theology." In *Naturalists, Explorers and Field Scientists in South-East Asia and Australia*, edited by Indraneil Das and Andrew Alek Tuens, 51–70. Dordrecht: Springer, 2016.

Franklin, J. Jeffrey. *Spirit Matters: Occult Beliefs, Alternative Religions, and the Crisis of Faith in Victorian England*. Ithaca, NY: Cornell University Press, 2018.

Garnett, Ronald George. *Co-Operation and the Owenite Socialist Communities in Britain, 1825–45*. Manchester: Manchester University Press, 1971.

Gasparin, Agénor de. *Science vs. Modern Spiritualism: A Treatise on Turning Tables, the Supernatural in General, and Spirits*. New York: Kiggins & Kellogg, 1856.

Gauld, Alan. *The Founders of Psychical Research*. London: Routledge and Kegan Paul, 1968.

Geertz, Clifford. *The Interpretation of Cultures*. New York: Basic Books, 1973.

Gieryn, Thomas F. *Cultural Boundaries of Science: Credibility on the Line*. Chicago: University of Chicago Press, 1999.

Gillespie, Neal C. *Charles Darwin and the Problem of Creation*. Chicago: University of Chicago Press, 1979.

Gilmour, Robin. *The Victorian Period: The Intellectual and Cultural Context of English Literature, 1830–1890*. London: Routledge, 1993.

Ginzburg, Carlo. *The Cheese and the Worms: The Cosmos of a Sixteenth-Century Miller*. Baltimore, MD: Johns Hopkins University Press, 1980.

Ginzburg, Carlo. "Microhistory: Two or Three Things That I Know about It." *Critical Inquiry* 20 (1993): 10–35.

[Gloumeline, Julie de.] *D. D. Home: His Life and Mission*. London: Trübner, 1888.

Golan, Tal. *Laws of Men and Laws of Nature: The History of Scientific Expert Testimony in England and America*. Cambridge, MA: Harvard University Press, 2004.

Goldsmith, Oliver. *The Mystery Revealed; Containing a Series of Transactions and Authentic Testimonials: Respecting the Supposed Cock-Lane Ghost: which have hitherto been Concealed from the Public*. London: W. Bristow, 1762.

Gooday, Graeme. *The Morals of Measurement: Accuracy, Irony and Trust in Late Victorian Electrical Practice*. Cambridge: Cambridge University Press, 2004.

Gordon, Margaret Maria. *The Home Life of Sir David Brewster*. Edinburgh: Edmonston and Douglas, 1869.

Grant, Douglas. *The Cock Lane Ghost*. London: Macmillan, 1965.

Grasseni, Cristina. Introduction to *Skilled Visions: Between Apprenticeship and Standards*, edited by Cristina Grasseni, 1–19. Oxford: Berghahn Books, 2007.

Grasseni, Cristina, ed. *Skilled Visions: Between Apprenticeship and Standards*. Oxford: Berghahn Books, 2007.

Green, Roger Lancelyn. *Andrew Lang: A Critical Biography with a Short-Title Bibliography of the Works of Andrew Lang*. Bloomington: Indiana University Press, 1946.

Gregory, Frederick. *Scientific Materialism in Nineteenth-Century Germany*. Dordrecht: Springer, 1977.

Grimes, Hilary. *The Late Victorian Gothic: Mental Science, the Uncanny, and Scenes of Writing*. Farnham: Ashgate, 2011.

Grimshaw, Anna. *The Ethnographer's Eye: Ways of Seeing in Modern Anthropology*. Cambridge: Cambridge University Press, 2001.

Gruber, Howard. *Darwin on Man: A Psychological Study of Scientific Creativity*. Chicago: University of Chicago Press, 1974.

Gupta, Akhil, and James Ferguson. "Discipline and Practice: 'The Field' as Site, Method, and Location in Anthropology." In *Anthropological Locations: Boundaries and Grounds of a Field Science*, edited by Akhil Gupta and James Ferguson, 1–46. Berkeley: University of California Press, 1997.

Gurney, Edmund, and Frederic W. H. Myers. "Transferred Impressions and Telepathy." *Fortnightly Review* 39 (1883): 437–52.

Gurney, Edmund, Frank Podmore, and Frederic W. H. Myers. *Phantasms of the Living*. 2 vols. London: Trübner, 1886.

Hacking, Ian. *The Social Construction of What?* Cambridge, MA: Harvard University Press, 1999.

Hall, David. *Witch-Hunting in Seventeenth-Century New England: A Documentary History, 1638–1693*. Durham, NC: Duke University Press, 2000.

Hall, Trevor H. *The Spiritualists: The Story of Florence Cook and William Crookes*. London: Duckworth, 1962.

Hall, Trevor H. *The Strange Case of Ada Goodrich-Freer*. London: Duckworth, 1968.

Hall, Trevor H. *The Strange Case of Edmund Gurney*. London: Duckworth, 1964.

Hallowell, Irving A. "Ojibwa Ontology, Behavior, and World View." In *Cultures in History: Essays in Honor of Paul Radin*, edited by Stanley Diamond, 17–49. New York: Columbia University Press, 1960.

Hamer, Felicity Tsering Chödron. "Helen F. Stuart and Hannah Frances Green: The Original Spirit Photographer." *History of Photography* 42, no. 2 (2018): 146–67.

Harrison, John F. C. "'The Steam Engine of the New Moral World': Owenism and Education, 1817–1829." *Journal of British Studies* 6, no. 2 (1967): 76–98.

Harrison, Peter. "'Science' and 'Religion': Constructing the Boundaries." *Journal of Religion* 86 (2006): 81–106.

Harrison, Peter. *The Territories of Science and Religion*. Chicago: University of Chicago Press, 2015.

Harvey, Graham. *Animism: Respecting the Living World*. London: Hurst, 2005.

Harvey, John. *Photography and Spirit*. London: Reaktion Books, 2007.

Haynes, Renée. *The Society for Psychical Research, 1882–1982*. London: Macdonald, 1982.

Hazelgrove, Jenny. "Spiritualism after the Great War." *Twentieth Century British History* 10, no. 4 (1999): 404–30.

Hazelgrove, Jenny. *Spiritualism and British Society between the Wars*. Manchester: Manchester University Press, 2000.

Headrick, Daniel. *Power over Peoples: Technology, Environments, and Western Imperialism, 1400 to the Present*. Princeton, NJ: Princeton University Press, 2009.

Headrick, Daniel. *The Tools of Empire: Technology and European Imperialism in the Nineteenth Century*. Oxford: Oxford University Press, 1981.

Helmstadter, Richard, and Bernard Lightman, eds. *Victorian Faith in Crisis: Essays on Continuity and Change in Nineteenth-Century Religious Belief*. Redwood City, CA: Stanford University Press, 1990.

Hesketh, Ian. *Of Apes and Ancestors: Evolution, Christianity, and the Oxford Debate*. Toronto: University of Toronto Press, 2009.

Hesketh, Ian. *Victorian Jesus: J. R. Seeley, Religion, and the Cultural Significance of Anonymity*. Toronto: University of Toronto Press, 2017.

Hoare, Philip. *England's Lost Eden: Adventures in Victorian Utopia*. London: Fourth Estate, 2005.

Holloway, Julian. "Enchanted Spaces: The Séance, Affect, and Geographies of Religion." *Annals of the Association of American Geographies* 96, no. 1 (2006): 182–87.

Houdini, Harry. *A Magician among the Spirits*. London: Harper & Brothers, 1924.

Houdini, Harry. "Spiritualism and Magic." Harry Houdini Papers. Harry Ransom Center, University of Texas at Austin.

Houghton, Georgiana. *Chronicles of the Photographs of Spiritual Beings*. London: E. W. Allen, 1882.

Howitt, William. *The History of the Supernatural in All Ages and Nations, and in All Churches, Christian and Pagan, Demonstrating a Universal Faith*. London: Longman, 1863.

Hulme, Peter, and Tim Youngs. Introduction to *The Cambridge Companion to Travel Writing*, edited by Peter Hulme and Tim Youngs, 1–16. Cambridge: Cambridge University Press, 2002.

Hunt, James. "Introductory Address on the Study of Anthropology." *Anthropological Review* 1 (1863): 1–20.

Hunt, James. "On Physio-anthropology, Its Aim and Method." *Journal of the Anthropological Society of London* 5 (1867): ccix–cclxxi.

Hunt, James. *On the Negro's Place in Nature*. London: Trübner, 1863.

Huxley, Thomas Henry. *Evidence as to Man's Place in Nature*. London: Williams and Norgate, 1863.

Huxley, Thomas Henry. *On Our Knowledge of the Causes of Organic Nature: Being Six Lectures to Working Men, Delivered at the Museum of Practical Geology*. London: Robert Hardwicke, 1863.

Huxley, Thomas Henry. "On the Methods and Results of Ethnology." *Fortnightly Review* 1 (1865): 257–77.

Isichei, Elizabeth. *Victorian Quakers*. Oxford: Oxford University Press, 1970.

Johnstone, M. E. "The Fox Sisters." Catalogue number 2009.148.2. Pitt Rivers Museum Manuscript Collection. University of Oxford.

Jolly, Martyn. *Faces of the Living Dead: The Belief in Spirit Photography*. London: Mark Batty, 2006.

Jones, Darryl. "'Gone into Mourning . . . for the Death of the Sun.'" In *Victorian Time: Technologies, Standardizations, Catastrophes*, edited by Trish Ferguson, 178–95. Basingstoke: Palgrave Macmillan, 2013.

Jones, Greta. "Alfred Russel Wallace, Robert Owen and the Theory of Natural Selection." *British Journal for the History of Science* 35 (2002): 73–96.

Josephson-Storm, Jason Ā. *The Myth of Disenchantment: Magic, Modernity, and the Birth of the Human Sciences*. Chicago: University of Chicago Press, 2017.

Jowett, Benjamin. "On the Interpretation of Scripture." In *Essays and Reviews*, edited by John William Parker, 330–433. London: John Parker and Son, 1860.

Kaalund, Nanna Katrine Lüders. "From Science in the Arctic to Arctic Science: A Transnational Study of Arctic Travel Narratives, 1818–1883." PhD diss., York University, Toronto, 2017.

Kaalund, Nanna Katrine Lüders. "Oxford Serialized: Revisiting the Huxley-Wilberforce Debate through the Periodical Press." *History of Science* 52, no. 4 (2014): 429–53.

Kalush, William, and Larry Sloman. *The Secret Life of Houdini: The Making of America's First Superhero*. New York: Simon and Schuster, 2006.

Kaplan, Louis. *The Strange Case of William Mumler, Spirit Photographer*. Minneapolis: University of Minnesota Press, 2008.

Kaplan, Louis. "Where the Paranoid Meets the Paranormal: Speculations on Spirit Photography." *Art Journal* 62, no. 3 (2003): 18–27.

Keezer, William. "Alfred Russel Wallace: Naturalist, Zoogeographer, Spiritualist, and Evolutionist." *Bios* 36, no. 2 (1965): 66–70.

Kenny, Robert. "From the Curse of Ham to the Curse of Nature: The Influence of Natural Selection on the Debate on Human Unity before the Publication of *The Descent of Man*." *British Journal for the History of Science* 40, no. 3 (2007): 367–88.

Kirchhoff, Gustav, and Robert Bunsen. "Chemical Analysis by Observation of Spectra." *Annalen der Physik und der Chemie* 10 (1860): 161–89.

Koerner, Lisbet. "Purposes of Linnaean Travel: A Preliminary Research Report." In *Visions of Empire: Voyages, Botany, and Representation in Nature*, edited by David Phillip Miller and Peter Hanns Reill, 117–52. Cambridge: Cambridge University Press, 1996.

Kottler, Malcolm Jay. "Alfred Russel Wallace, the Origin of Man, and Spiritualism." *Isis* 65 (1974): 144–92.

Krage, Helge S. *Entropic Creation: Religious Contexts of Thermodynamics and Cosmology*. London: Routledge, 2016.

Kuklick, Henrika. "After Ishmael: The Fieldwork Tradition and Its Future." In *Anthropological Locations: Boundaries and Grounds of a Field Site*, edited by Akhil Gupta and James Ferguson, 47–65. Berkeley: University of California Press, 1997.

Kuklick, Henrika. "The British Tradition." In *A New History of Anthropology*, edited by Henrika Kuklick, 52–78. Oxford: Blackwell, 2008.

Kuklick, Henrika. Introduction to *A New History of Anthropology*, edited by Henrika Kuklick, 1–16. Oxford: Blackwell, 2008.

Kuklick, Henrika, ed. *A New History of Anthropology*. Oxford: Blackwell, 2008.

Kuklick, Henrika. "Personal Equations: Reflections on the History of Fieldwork, with Special Reference to Sociocultural Anthropology." *Isis* 102 (2011): 1–33.

Kuklick, Henrika. *The Savage Within: The Social History of British Anthropology, 1885–1945*. Cambridge: Cambridge University Press, 1991.

Kuklick, Henrika. "The Theory of Evolution and Cultural Anthropology." In *The Theory of Evolution and Its Impact*, edited by Aldo Fasolo, 83–102. Berlin: Springer, 2012.

Kuper, Adam. *Anthropology and Anthropologists: The Modern British School*. London: Routledge, 1993.

Kurtz, Paul. *A Skeptic's Handbook of Parapsychology*. Buffalo, NY: Prometheus Books, 1985.

Laidlaw, Zoë. *Colonial Connections, 1815–45: Patronage, the Information Revolution and Colonial Government*. Manchester: Manchester University Press, 2005.

Laidlaw, Zoë. "Heathens, Slaves and Aborigines: Thomas Hodgkin's Critique of Missions and Anti-slavery," *History Workshop Journal* 64, no. 1 (2007): 134–61.

Lamb, Geoffrey. *Victorian Magic*. Abingdon, UK: Routledge, 1976.

Lamont, Peter. *Extraordinary Beliefs: A Historical Approach to a Psychological Problem*. Cambridge: Cambridge University Press, 2013.

Lamont, Peter. *The First Psychic: The Peculiar Mystery of a Notorious Victorian Wizard*. London: Abacus Books, 2005.

Lamont, Peter. "Magician as Conjuror: A Frame Analysis of Victorian Mediums." *Early Popular Visual Culture* 4, no. 1 (2006): 21–33.

Lamont, Peter. "Spiritualism and a Mid-Victorian Crisis of Evidence." *Historical Journal* 47, no. 4 (2004): 897–920.

Lane, Christopher. *The Age of Doubt: Tracing the Roots of Our Religious Uncertainty*. New Haven, CT: Yale University Press, 2011.

Lang, Andrew. *Cock Lane and Common Sense*. London: Longmans, Green, 1894.

Lang, Andrew. "The Comparative Study of Ghost Stories." *Nineteenth Century* 17 (1885): 623–32.

Lang, Andrew. *Custom and Myth*. London: Longmans, Green, 1884.

Lang, Andrew, "Edward Burnett Tylor." In *Anthropological Essays Presented to Edward Tylor in Honour of his 75th Birthday*, edited by Henry Balfour, 1–17. Oxford: Clarendon, 1907.

Lang, Andrew. *Myth, Ritual and Religion*. London: Longmans, Green, 1887.

Lang, Andrew. "The Poltergeist at Cideville." *Proceedings of the Society for Psychical Research* 18 (1903–1904): 454–63.

Lang, Andrew. "Protest of a Psycho-Folklorist." *Folklore* 6 (1895): 236–48.

Larsen, Timothy. *Crisis of Doubt: Honest Faith in Nineteenth-Century England*. Oxford: Oxford University Press, 2006.

Larsen, Timothy. *The Slain God: Anthropologists and the Christian Faith*. Oxford: Oxford University Press, 2014.

Latham, Robert Gordon. *The Natural History of the Varieties of Man*. London: John van Voorst, 1850.

Lawrence, William. *Lectures on Physiology, Zoology, and the Natural History of Man, Delivered at the Royal College of Surgeons*. Salem: Foote and Brown, 1828.

Lecourt, Sebastian. *Cultivating Belief: Victorian Anthropology, Liberal Aesthetics, and the Secular Imagination*. Oxford: Oxford University Press, 2018.

Lehman, Amy. *Victorian Women and the Theatre of Trance: Mediums, Spiritualists and Mesmerists in Performance*. London: MacFarland, 2002.

Leopold, Joan. *Culture in Comparative and Evolutionary Perspective: E .B. Tylor and the Making of Primitive Culture*. Berlin: Dietrich Reimer, 1980.

Lepore, Jill. "Historians Who Love Too Much: Reflections on Microhistory and Biography." *Journal of American History* 88, no. 1 (2001): 129–44.

Lester, Joseph. *E. Ray Lankester and the Making of Modern Biology*. Farringdon: British Society for the History of Science, 1995.

Levi, Giovanni. "On Microhistory." In *New Perspectives on Historical Writing*, edited by Peter Burke, 93–113. Cambridge: Polity, 1992.

Lightman, Bernard. "Darwin and the Popularization of Evolution." *Notes and Records of the Royal Society of London* 64 (2010): 5–24.

Lightman, Bernard. "Does the History of Science and Religion Change Depending on the Narrator? Some Atheist and Agnostic Perspectives." *Science and Christian Beliefs* 24 (2012): 149–68.

Lightman, Bernard. "Huxley and Scientific Agnosticism: The Strange History of a Failed Rhetorical Strategy." *British Journal for the History of Science* 35 (2002): 271–89.

Lightman, Bernard. *The Origins of Agnosticism: Victorian Unbelief and the Limits of Knowledge*. Baltimore, MD: Johns Hopkins University Press, 1987.

Lightman, Bernard. *Victorian Popularizers of Science: Designing Nature for New Audiences*. Chicago: University of Chicago Press, 2007.

Lightman, Bernard. "Victorian Sciences and Religion: Discordant Harmonies." *Osiris* 16 (2001): 343–66.

Lodge, Oliver. *Raymond, or, Life and Death: With Examples of the Evidence for Survival of Memory and Affection after Death*. London: Methuen, 1916.

Lorimer, Douglas A. "Science and the Secularization of Victorian Images of Race." In *Victorian Science in Context*, edited by Bernard Lightman, 212–35. Chicago: University of Chicago Press, 1997.

Lorimer, Douglas A. *Science, Race Relations and Resistance: Britain, 1870–1914*. Manchester: Manchester University Press, 2013.

Lowrey, Kathleen Bolling. "Alfred Russel Wallace as Ancestor Figure: Reflections on Anthropological Lineage after the Darwin Bicentennial." *Anthropology Today* 26 (2010): 18–21.

Luckhurst, Roger. *The Invention of Telepathy, 1870–1901.* Oxford: Oxford University Press, 2002.

Luckhurst, Roger. "Knowledge, Belief and the Supernatural at the Imperial Margin." In *The Victorian Supernatural,* edited by Nicola Bown, Carolyn Burdett, and Pamela Thurschwell, 197–216. Cambridge: Cambridge University Press, 2004.

Luckhurst, Roger. "An Occult Gazetteer of Bloomsbury: An Experiment in Method." In *London Gothic: Place, Space and the Gothic Imagination,* edited by Lawrence Phillips and Anne Veronica Witchard, 50–64. London: Continuum Literary Studies, 2010.

Lyons, Sherrie Lynne. *Species, Serpents, Spirits and Skulls: Science at the Margins in the Victorian Age.* Albany: State University of New York Press, 2009.

Mackay, Charles. *Memoirs of Extraordinary Popular Delusions.* 2nd ed. 2 vols. London: Office of the National Illustrated Library, 1852.

MacLeod, Roy M. "The X Club: A Social Network of Science in Late-Victorian England." *Notes and Records of the Royal Society of London* 24 (1970): 305–22.

Magnússon, Sigurður Gylfi. "Far-reaching Microhistory: The Use of Microhistorical Perspective in a Globalized World." *Rethinking History* 21, no. 3 (2017): 312–41.

Magnússon, Sigurður Gylfi. "Microhistory, Biography and Ego-Documents in Historical Writing." *Revue d'histoire nordique* 20 (2016): 133–53.

Magnússon, Sigurður Gylfi. "'The Singularization of History': Social History and Microhistory within the Postmodern State of Knowledge." *Journal of Social History* 36, no. 3 (2003): 701–35.

Magnússon, Sigurður Gylfi. "Social History as 'Sites of Memory'? The Institutionalization of History: Microhistory and the Grand Narrative." *Journal of Social History* 39, no. 3 (2006): 891–913.

Magnússon, Sigurður Gylfi, and István M. Szjártó. *What Is Microhistory? Theory and Practice.* Abingdon, UK: Routledge, 2013.

Manias, Chris. "The Problematic Construction of 'Paleolithic Man': The Old Stone Age and the Difficulties of the Comparative Method, 1859–1914." *Studies in History and Philosophy of Biological and Biomedical Sciences* 51 (2015): 32–43.

Manias, Chris. *Race, Science, and the Nation: Reconstructing the Ancient Past in Britain, France and Germany.* Abingdon, UK: Routledge, 2013.

Marchant, James, ed. *Alfred Russel Wallace: Letters and Reminiscences.* 2 vols. New York: Cassell, 1916.

Marett, Robert Ranulph. *Tylor.* London: Chapman and Hall, 1936.

Marsden, Ben, and Crosbie Smith. *Engineering Empires: A Cultural History of Technology in Nineteenth-Century Britain.* Basingstoke: Palgrave Macmillan, 2005.

Mather, Cotton. *Wonders of the Invisible World: Observations As well Historical as Theological, upon the Nature, the Number, and the Operations of the Devils.* Boston: Benjamin Harris, 1693.

Mauskopf, Seymour, and Michael R. McVaugh. *The Elusive Science: Origins of Experimental Psychical Research.* Baltimore, MD: Johns Hopkins University Press, 1980.

McCabe, Joseph. *Edward Clodd: A Memoir.* London: John Lane, 1932.

McCabe, Joseph. *Spiritualism: A Popular History from 1847.* New York: Dodd, Mead, 1920.

McCorristine, Shane. *The Spectral Arctic: A History of Ghosts and Dreams in Polar Exploration.* London: University College London Press, 2018.

McCorristine, Shane. *Spectres of the Self: Thinking about Ghosts and Ghost-Seeing in England, 1750–1920.* Cambridge: Cambridge University Press, 2010.

McLennan, John Ferguson. *Primitive Marriage: An Inquiry into the Origin of the Form of Capture in Marriage Ceremonies*. Edinburgh: Adam and Charles Black, 1865.

Meade, Marion. *Madame Blavatsky: The Woman behind the Myth*. New York: Putnam, 1980.

"Michael Faraday's Researches in Spiritualism." *Scientific Monthly* 83 (1956): 145–50.

Mifflin, Jeffrey. "Visual Archives in Perspective: Enlarging on Historical Medical Photographs." *American Archivist* 70, no. 1 (2007): 32–69.

Mirville, Charles-Jules de. *Fragment d'un Ouvrage Inédit*. Paris: A. Chérié, 1852.

Mitchell, Benjamin David. "Capturing the Will: Imposture, Delusion, and Exposure in Alfred Russel Wallace's Defense of Spirit Photography." *Studies in History and Philosophy of Biological and Biomedical Sciences* 46 (2014): 15–24.

Moore, James R. *The Post-Darwinian Controversies: A Study of the Protestant Struggle to Come to Terms with Darwin in Great Britain and America, 1870–1900*. Cambridge: Cambridge University Press, 1979.

Morus, Iwan Rhys. "Illuminating Illusions, or, the Victorian Art of Seeing Things." *Early Popular Visual Culture* 10, no. 1 (2012): 37–50.

Moses, William Stainton. *Direct Spirit Writing (Psychography): A Treatise on One of the Objective Forms of Psychic or Spiritual Phenomena*. London: Psychic Book Club, 1878.

Moses, William Stainton. *Spirit Teachings through the Mediumship of William Stainton Moses*. London: Spiritualist Alliance, 1883.

"Mrs. Jennie Holmes's Séances." *Medium and Daybreak* 3 (1872): 436.

"Mrs. Olive, Trance Medium." *Medium and Daybreak* 3 (1872): 436.

Mullin, Rita T. *Harry Houdini: Death-Defying Showman*. Cambridge: Baker and Taylor, 2009.

Mumler, William Howard. "Charles H. Foster." Albumen silver print, photographed in Boston, MA, between 1862 and 1875. Catalogue number 84.XD.760.1.9. J. Paul Getty Museum, Los Angeles, CA.

Murphet, Howard. *When Daylight Comes: A Biography of Helena Petrovna Blavatsky*. Wheaton, IL: Theosophical Publishing House, 1975.

Murphree, Idus L. "The Evolutionary Anthropologists: The Progress of Mankind: The Concepts of Progress and Culture in the Thought of John Lubbock, Edward B. Tylor, and Lewis H. Morgan." *Proceedings of the American Philosophical Society* 105, no. 3 (1961): 265- 300.

Murray, Gilbert. *Four Stages of Greek Religion: Studies Based on a Course of Lectures Delivered in April 1912 at Columbia University*. New York: Columbia University Press, 1912.

Myers, Frederic W. H. "On a Telepathic Explanation of Some So-Called Spiritualistic Phenomena." *Proceedings of the Society for Psychical Research* 2 (1884): 217–37.

Myers, Frederic W. H. "On Telepathic Hypnotism, and Its Relation to Other Forms of Hypnotic Suggestion." *Proceedings of the Society for Psychical Research* 4 (1886): 127–88.

Myers, Frederic W. H. "The Subliminal Consciousness." *Proceedings of the Society for Psychical Research* 7 (1891–1892): 298–355.

Myers, Frederic W. H. "The Subliminal Consciousness." *Proceedings of the Society for Psychical Research* 9 (1893–1894): 3–128.

Nartonis, David. "The Rise of Nineteenth-Century American Spiritualism, 1854–1873." *Journal for the Scientific Study of Religion* 49, no. 2 (2010): 361–73.

Natale, Simone. "A Short History of Superimposition: From Spirit Photography to Early Cinema." *Early Popular Visual Culture* 10, no. 2 (2012): 125–45.

Natale, Simone. *Supernatural Entertainments: Victorian Spiritualism and the Rise of Modern Media Culture*. University Park: Pennsylvania State University Press, 2016.

Nelson, Geoffrey. *Spiritualism and Society*. Abingdon, UK: Routledge, 1969.

Nichols, Thomas Low. *A Biography of the Brothers Davenport: With Some Account of the Physical and Psychical Phenomena which Have Occurred in Their Presence, in America and Europe*. London: Sounders, Otley, 1864.

Noakes, Richard. "Haunted Thoughts of the Careful Experimentalist: Psychical Research and the Troubles of Experimental Physics." *Studies in History and Philosophy of Biological and Biomedical Sciences* 48 (2014): 46–56.

Noakes, Richard. *Physics and Psychics: The Occult and the Sciences in Modern Britain*. Cambridge: Cambridge University Press, 2019.

Noakes, Richard. "Spiritualism, Science and the Supernatural in Mid-Victorian Britain." In *The Victorian Supernatural*, edited by Nicola Bown, Carolyn Burdett, and Pamela Thurschwell, 23–43. Cambridge: Cambridge University Press, 2004.

Noyes, Deborah. *The Magician and the Spirits: Harry Houdini and the Curious Pastime of Communicating with the Dead*. New York: Penguin Books, 2017.

Oppenheim, Janet. *The Other World: Spiritualism and Psychical Research in England, 1850–1914*. Cambridge: Cambridge University Press, 1985.

Owen, Alex. *The Darkened Room: Women, Power, and Spiritualism in Late Victorian England*. London: Virago, 1989.

Owen, Robert. *The Future of the Human Race; or Great, Glorious, and Peaceful Revolution to be Effected through the Agency of Departed Spirits of Good and Superior Men and Women*. London: Effingham Wilson, 1854.

Owen, Robert Dale. *The Debatable Land between This World and the Next*. New York: G. W. Carleton, 1871.

Owen, Robert Dale. *Footfalls on the Boundary of Another World*. Philadelphia: J. B. Lippincott, 1860.

Pals, Daniel L. *The Victorian "Lives" of Jesus*. San Antonio, TX: Trinity University Press, 1982.

Pels, Peter. "Spiritual Facts and Super-Visions: The 'Conversion' of Alfred Russel Wallace." *Religion and Modernity* 8, no. 2 (1995): 69–91.

Podmore, Frank. *Apparitions and Thought-Transference: An Examination of the Evidence for Telepathy*. London: Walter Scott, 1894.

Podmore, Frank. *Modern Spiritualism: A History and a Criticism*. 2 vols. London: Methuen, 1902.

Podmore, Frank. *Robert Owen: A Biography*. 2 vols. London: Hutchinson, 1906.

Porche, Jean, and Deborah Vaughan. *Psychics and Mediums in Canada*. Toronto: Dundurn Group, 2005.

Porter, Roy. *The Making of Geology: Earth Science in Britain, 1660–1815*. Cambridge: Cambridge University Press, 1977.

Pratt, Mary Louise. *Imperial Eyes: Travel Writing and Transculturation*. London: Routledge, 1992.

Prichard, James Cowles. *Researches into the Physical History of Man*. London: John and Arthur Arch, 1813.

Pritchard, Linda K. "The Burned-Over District Reconsidered: A Portent of Evolving Religious Pluralism in the United States." *Social Science History* 8, no. 3 (1984): 243–65.

Quiggin, A. Hingston. *Haddon, the Head Hunter: A Short Sketch of the Life of A. C. Haddon*. Cambridge: Cambridge University Press, 1942.

Qureshi, Sadiah. *Peoples on Parade: Exhibitions, Empire, and Anthropology in Nineteenth-Century Britain.* Chicago: University of Chicago Press, 2011.

Raia, Courtenay. *The New Prometheans: Faith, Science, and the Supernatural Mind in the Victorian Fin de Siècle.* Chicago: University of Chicago Press, 2019.

Ratnapalan, Laavanyan. "E. B. Tylor and the Problem of Primitive Culture." *History and Anthropology* 19 (2008): 131–42.

Reinach, Salomon. "Andrew Lang." *Quarterly Review* 218 (1913): 309–19.

Richards, Evelleen. "The 'Moral Anatomy' of Robert Knox: The Interplay between Biological and Social Thought in Victorian Scientific Naturalism." *Journal of the History of Biology* 22 (1989): 373–436.

Richardson, Elsa. *Second Sight in the Nineteenth Century: Prophecy, Imagination and Nationhood.* Basingstoke: Palgrave Macmillan, 2017.

Rinn, Joseph Francis. *Sixty Years of Psychical Research: Houdini and I among the Spiritualists.* New York: Truth Seeker, 1950.

Roach, Mary. *Spook: Science Tackles the Afterlife.* New York: W. W. Norton, 2005.

Robertson, Beth A. *Science of the Séance: Transnational Networks and Gendered Bodies in the Study of Psychic Phenomena, 1918–1940.* Vancouver: University of British Columbia Press, 2016.

Rosenthal, Bernard. *Salem Story: Reading the Witch Trials of 1692.* Cambridge: Cambridge University Press, 1993.

Rudwick, Martin. *The Great Devonian Controversy: The Shaping of Scientific Knowledge among Gentlemanly Specialists.* Chicago: University of Chicago Press, 1985.

Salter, William Henry. *The Society for Psychical Research: An Outline of Its History.* London: Society for Psychical Research, 1948.

Sandford, Christopher. *Houdini and Conan Doyle: The Great Magician and the Inventor of Sherlock Holmes. Friends of Genius, Deadly Rivals.* London: Duckworth Overlook, 2011.

Savory, Tanya. *The Amazing Harry Houdini.* West Berlin, NJ: Townsend, 2009.

Schaffer, Simon. "Astronomers Mark Time: Discipline and the Personal Equation." *Science in Context* 2, no. 1 (1988): 115–45.

Secord, Anne. "Artisan Naturalists: Science as Popular Culture in Nineteenth-Century England." PhD diss., University of London, 2002.

Secord, Anne. "Corresponding Interests: Artisans and Gentlemen in Nineteenth-Century Natural History." *British Journal for the History of Science* 27 (1994): 383–408.

[Seeley, John Robert]. *Ecce Homo: A Survey in the Life and Work of Jesus Christ.* London: Macmillan, 1865.

Sera-Shriar, Efram. "Anthropometric Portraiture and Victorian Anthropology: Situating Francis Galton's Photographic Work in the Late 1870s." *History of Science* 53, no. 2 (2015): 155–79.

Sera-Shriar, Efram. "Arctic Observers: Richard King, Monogenism and the Historicisation of Inuit through Travel Narratives." *Studies in History and Philosophy of Biological and Biomedical Sciences* 51 (2015): 23–31.

Sera-Shriar, Efram. "Credible Witnessing: A. R. Wallace, Spiritualism, and a 'New Branch of Anthropology.'" *Modern Intellectual History* 17, no. 2 (2020): 357–84.

Sera-Shriar, Efram. "From the Beginning: Human History Theories in Nineteenth-Century British Sciences." In *Historicizing Humans: Deep Time, Evolution, and Race in Nineteenth-Century British Sciences,* edited by Efram Sera-Shriar, 1–13. Pittsburgh: University of Pittsburgh Press, 2018.

Sera-Shriar, Efram. "Historicizing Belief: E. B. Tylor, *Primitive Culture*, and the Evolution of Religion." In *Historicizing Humans: Deep Time, Evolution and Race in Nineteenth-Century British Sciences*, edited by Efram Sera-Shriar, 68–90. Pittsburgh: University of Pittsburgh Press, 2018.

Sera-Shriar, Efram, ed. *Historicizing Humans: Deep Time, Evolution, and Race in Nineteenth-Century British Sciences*. Pittsburgh: University of Pittsburgh Press, 2018.

Sera-Shriar, Efram. "Human History and Deep Time in Nineteenth-Century British Sciences: An Introduction." *Studies in History and Philosophy of Biological and Biomedical Sciences* 51 (2015): 19–22.

Sera-Shriar, Efram. *The Making of British Anthropology, 1813–1871*. London: Pickering & Chatto, 2013.

Sera-Shriar, Efram. "Observing Human Difference: James Hunt, Thomas Huxley, and Competing Disciplinary Strategies in the 1860s." *Annals of Science* 70, no. 4 (2013): 461–91.

Sera-Shriar, Efram. "Race." In *Historicism and the Human Sciences in Victorian Britain*, edited by Mark Bevir, 48–76. Cambridge: Cambridge University Press, 2017.

Sera-Shriar, Efram. "Tales from Patagonia: Phillip Parker King and Early Ethnographic Observation in British Ethnology, 1826–1830." *Studies in Travel Writing* 19, no. 3 (2015): 204–23.

Sera-Shriar, Efram. "What Is Armchair Anthropology? Observational Practices in Nineteenth-Century British Human Sciences." *History of the Human Sciences* 27, no. 2 (2014): 180–94.

[Sexton, S. S. P.] "The True Portrait of the Ghost." In *The Beauties of All Magazines Selected for 1762*, edited by George Alexander Stevens, vol. 1, 48. London: T. Waller, 1762.

Shapin, Steven. *A Social History of Truth: Civility and Science in Seventeenth-Century England*. Chicago: University of Chicago Press, 1994.

Shapin, Steven. *Never Pure: Historical Studies of Science as If It Was Produced by People with Bodies Situated in Time, Space, Culture, and Society, and Struggling for Credibility and Authority*. Baltimore, MD: Johns Hopkins University Press, 2010.

Shapin, Steven. *The Scientific Life: A Moral History of a Late Modern Vocation*. Chicago: University of Chicago Press, 2008.

Shapin, Steven, and Simon Schaffer. *Leviathan and the Air-Pump: Hobbes, Boyle, and the Experimental Life*. Princeton, NJ: Princeton University Press, 1985.

Shea, Victor, and William Whitla, eds. *Essays and Reviews: The 1860 Text and Its Reading*. Charlottesville: University of Virginia Press, 2000.

Shermer, Michael. *In Darwin's Shadow: The Life and Science of Alfred Russel Wallace*. Oxford: Oxford University Press, 2002.

Schlossberg, Herbert. *Conflict and Crisis in the Religious Life of Late Victorian England*. New Brunswick, NJ: Transaction, 2009.

Sidgwick, Henry, et al. "Report on the Census of Hallucinations." *Proceedings of the Society for Psychical Research* 10 (1894): 25–422.

Slotten, Ross. *The Heretic in Darwin's Court: The Life of Alfred Russel Wallace*. New York: Columbia University Press, 2004.

Smith, Crosbie. *The Science of Energy: A Cultural History of Energy Physics in Victorian Britain*. Chicago: University of Chicago Press, 1998.

Sommer, Andreas. "Psychical Research in the History and Philosophy of Science: An Introduction and Review." *Studies in History and Philosophy of Biological and Biomedical Sciences* 48 (2014): 38–45.

"Sorcellerie: Relation de l'Événement du Presbytère de Cideville." *La Table Parlante: Journal des Faits Merveilleux* 1 (1854–1856): 129–73.

Spencer, Frank. "Hunt, James (1833–1869)." In *History of Physical Anthropology*, edited by Frank Spencer, 1:506–8. London: Garland, 1997.

Stanley, Matthew. *Huxley's Church and Maxwell's Demon: From Theistic Science to Naturalistic Science.* Chicago: University of Chicago Press, 2014.

Steinmeyer, Jim. *Hiding the Elephant: How Magicians Invented the Impossible and Learned to Disappear.* London: Arrow, 2005.

Stocking, George W., Jr. *After Tylor: British Social Anthropology, 1888–1951.* Madison: University of Wisconsin Press, 1996.

Stocking, George W., Jr. "Animism in Theory and Practice: E. B. Tylor's Unpublished 'Notes on "Spiritualism."'" *Man* n.s. 6, no. 1 (1971): 88–104.

Stocking, George W., Jr. "'Cultural Darwinism' and 'Philosophical Idealism' in E. B. Tylor." In *Race, Culture, and Evolution: Essays in the History of Anthropology*, edited by George W. Stocking Jr., 91–109. Chicago: University of Chicago Press, 1968.

Stocking, George W., Jr. "Edward Burnett Tylor and the Mission of Primitive Man." In *Delimiting Anthropology: Occasional Essays and Reflections.* edited by George W. Stocking Jr., 103–15. Madison: University of Wisconsin Press, 2001.

Stocking, George W., Jr. *The Ethnographer's Magic and Other Essays in the History of Anthropology.* Madison: University of Wisconsin Press, 1992.

Stocking, George W., Jr. "Matthew Arnold, E. B. Tylor, and the Uses of Invention." *American Anthropologist* 65, no. 4 (1963): 783–99.

Stocking, George W., Jr., ed. *Observers Observed: Essays on Ethnographic Fieldwork.* Madison: University of Wisconsin Press, 1983.

Stocking, George W., Jr. "On the Limits of 'Presentism' and 'Historicism' in the Historiography of the Behavioral Sciences." In *Race, Culture, and Evolution: Essays in the History of Anthropology*, edited by George W. Stocking Jr., 1–12. London: Blackwell, 1982.

Stocking, George W., Jr. *Race, Culture and Evolution: Essays in the History of Anthropology.* Chicago: University of Chicago Press, 1968.

Stocking, George W., Jr. *Victorian Anthropology.* New York: Free Press, 1987.

Stocking, George W., Jr. "What's in a Name? The Origins of the Royal Anthropological Institute (1837–71)." *Man* 6, no. 3 (1971): 369–90.

Strenski, Ivan. *Thinking about Religion: An Historical Introduction to Theories of Religion.* Oxford: Blackwell, 2006.

Stringer, Martin. "Rethinking Animism: Thoughts from the Infancy of Our Discipline." *Journal of the Royal Anthropological Institute* 5, no. 4 (1999): 541–55.

Sword, Helen. *Ghostwriting Modernism.* Ithaca, NY: Cornell University Press, 2002.

Taves, Ann. *Fits, Trances, and Visions: Experiencing Religion and Explaining Experience from Wesley to James.* Princeton, NJ: Princeton University Press, 1999.

Tedlock, Barbara. "From Participant Observation to the Observation of Participation: The Emergence of Narrative Ethnography." *Journal of Anthropological Research* 47, no. 1 (1991): 69–94.

Thompson, E. P. "Patrician Society, Plebeian Culture." *Journal of Social History* 7, no. 4 (1974): 382–405.

Thomson, William. "On a Universal Tendency in Nature to the Dissipation of Energy." *Philosophical Magazine* 4 (1852): 304–6.

Thomson, William. "On the Age of the Sun's Heat." *Macmillan's Magazine* 5 (1862): 388–93.

Thury, Marc. *Les Tables Tournantes: Considérées au Point de Vue de la Question de Physique Générale qui s'y Rattache, le Livre de M. le Comte A. de Gasparin et les Expériences de Valleyres.* Geneva: J. Kessman, 1855.

Thyraeus, Petrus. *Loca Infesta, Hoc Est: De Infestis, ob Molestantes Daemoniorvm et Defvnctorvm Hominvm Spiritvs, Locis, Liber Vnvs.* Cologne: Cholinus, 1598.

Tristano, Richard M. "Microhistory and Holy Family Parish: Some Methodological Considerations." *U.S. Catholic Historian* 14, no. 3 (1996): 23–30.

Truesdell, John W. *The Bottom Facts Concerning the Science of Spiritualism: Derived from Careful Investigation Covering a Period of Twenty-Five Years.* New York: G. W. Carleton, 1883.

Tuckett, Ivor Lloyd. *The Evidence for the Supernatural: A Critical Study Made with "Uncommon Sense".* London: Kegan Paul, Trench, Trübner, 1911.

Tucker, Jennifer. "The Historian, the Picture and the Archive." *Isis* 97, no. 1 (2006): 111–20.

Tucker, Jennifer. *Nature Exposed: Photography as Eyewitness in Victorian Science.* Baltimore, MD: Johns Hopkins University Press, 2005.

Tucker, Jennifer. "Photography as Witness, Detective, and Impostor: Visual Representation in Victorian Science." In *Victorian Science in Context*, edited by Bernard Lightman, 378–408. Chicago: University of Chicago Press, 1997.

Turner, Frank M. *Contesting Cultural Authority: Essays in Victorian Intellectual Life.* Cambridge: Cambridge University Press, 1993.

Turner, Frank M. "The Victorian Conflict between Science and Religion: A Professional Dimension." *Isis* 69 (1978): 356–76.

Tylor, Edward Burnett, to Alfred Russel Wallace, November 26, 1866. Add 46439 ff. 6. Tylor Papers. British Library, London.

Tylor, Edward Burnett. *Anahuac: or Mexico and the Mexicans, Ancient and Modern.* London: Longman, Green, Longman, and Roberts, 1861.

Tylor, Edward Burnett. *Anthropology: An Introduction to the Study of Man and Civilization.* London: Macmillan, 1881.

Tylor, Edward Burnett. Notebook on Spiritualism. Item 12, box 3. Pitt Rivers Museum Manuscript Collection. University of Oxford.

Tylor, Edward Burnett. "On the Survival of Savage Thought in Modern Civilization." *Notices of the Proceedings at the Meetings of the Members of the Royal Institution of Great Britain* 5 (1869): 522–35.

Tylor, Edward Burnett. *Primitive Culture: Researches into the Development of Mythology, Philosophy, Religion, Art and Custom.* 2 vols. London: John Murray, 1871.

Tylor, Edward Burnett. *Researches into the Early History of Mankind and the Development of Civilization.* London: John Murray, 1865.

Urry, James. "A History of Field Methods." In *Ethnographic Research: A Guide to General Conduct*, edited by Roy Frank Ellen, 27–61. London: Academic, 1984.

Urry, James. "Notes and Queries on Anthropology and the Development of Field Methods in British Anthropology, 1870–1920." *Proceedings of the Royal Anthropological Institute of Great Britain and Ireland* 1972 (1972): 45–57.

van Wyhe, John. *Dispelling the Darkness: Voyage in the Malay Archipelago and the Discovery of Evolution by Wallace and Darwin.* Singapore: World Scientific, 2013.

van Wyhe, John. Introduction to *The Annotated Malay Archipelago*, by Alfred Russel Wallace, edited by John van Wyhe, 1–38. Singapore: National University of Singapore Press, 2015.

Vargas-Cetina, Gabriela, ed. *Anthropology and the Politics of Representation*. Tuscaloosa: University of Alabama Press, 2013.

Vasconcelos, João. "Homeless Spirits: Modern Spiritualism, Psychical Research and the Anthropology of Religion in the Late Nineteenth and Early Twentieth Centuries." In *On the Margins of Religion*, edited by Frances Prin and João de Pina-Cabral, 13–37. Oxford: Berghahn Books, 2008.

Vetter, Jeremy. "The Unmaking of an Anthropologist: Wallace Returns from the Field, 1862–70." *Notes and Records of the Royal Society of London* 64 (2010): 25–42.

Visvanathan, Shiv. "Alternative Science." *Theory, Culture, and Society* 23, no. 2 (2006): 164–69.

Walkowitz, Judith. "Science and the Séance: Transgressions of Gender and Genre in Late Victorian London." *Representations* 22 (1988): 3–29.

Wallace, Alfred Russel, to Thomas Henry Huxley, November 22, 1866. Add. 46439 f. 5. Wallace Papers. British Library, London.

Wallace, Alfred Russel. "An Answer to the Arguments of Hume, Lecky, and Others, against Miracles." *Spiritualist: A Record of the Progress of the Science and Ethics of Spiritualism* 1 (1870): 113–16.

Wallace, Alfred Russel. "A Defense of Modern Spiritualism: Part I." *Fortnightly Review* 15 (1874): 630–57.

Wallace, Alfred Russel. "A Defense of Modern Spiritualism: Part II." *Fortnightly Review* 15 (1874): 785–807.

Wallace, Alfred Russel. *On Miracles and Modern Spiritualism: Three Essays*. London: James Burns, 1875.

Wallace, Alfred Russel. *My Life: A Record of Events and Opinions*. 2 vols. London: Chapman and Hall, 1905.

Wallace, Alfred Russel. "The Origin of Human Races and the Antiquity of Man Deduced from the Theory of 'Natural Selection.'" *Journal of the Anthropological Society of London* 2 (1864): clviii–clxxxvii.

Wallace, Alfred Russel. "Primitive Culture: Researches into the Development of Mythology, Philosophy, Religion, Art, and Custom." *Academy* 3 (1872): 69–71.

Wallace, Alfred Russel. *The Scientific Aspect of the Supernatural: Indicating the Desirableness of and Experimental Enquiry by Men of Science into the Alleged Powers of Clairvoyants and Mediums*. London: F. Farrah, 1866.

Wallace, Alfred Russel. "Spiritualism." In *Chambers's Encyclopedia*, edited by David Patrick, 9:645–49. Edinburgh: William and Robert Chambers, 1892.

Wallace, Alfred Russel. Spiritualism Journal. WCP5223.5749. Papers of Alfred Russel Wallace. Natural History Museum, London.

Warner, Marina. *Phantasmagoria: Spirit Visions, Metaphors, and Media into the Twenty-First Century*. Oxford: Oxford University Press, 2006.

Weaver, Janice. *Harry Houdini: The Legend of the World's Greatest Escape Artist*. New York: Harry N. Abrams, 2011.

Weisberg, Barbara. *Talking to the Dead: Kate and Maggie Fox and the Rise of Spiritualism*. San Francisco, CA: HarperCollins, 2004.

Wellman, Judith. *Grassroots Reform in the Burned-Over District of Upstate New York: Religion, Abolitionism, and Democracy*. Boca Raton, FL: Routledge, 2000.

Wheeler-Barclay, Marjorie. *The Science of Religion in Britain, 1860–1915*. Charlottesville: University of Virginia Press, 2010.

White, Paul. *Thomas Huxley: Making the "Man of Science"*. Cambridge: Cambridge University Press, 2003.

Wiley, Barry H. *The Thought Reader Craze: Victorian Science at the Enchanted Boundary*. London: McFarland, 2012.

Wilson, John G. *The Forgotten Naturalist: In Search of Alfred Russel Wallace*. Tennyson, Australia: Arcadia, 1999.

Winter, Alison. *Mesmerized: Powers of Mind in Victorian Britain*. Chicago: University of Chicago Press, 2000.

Whittington-Egan, Molly. *Mrs Guppy Takes a Flight: A Scandal of Victorian Spiritualism*. Castle Douglas, Scotland: Neil Wilson, 2014.

Wright, John, ed. *Letters of Horace Walpole*. 6 vols. London: R. Bentley, 1840.

Wulf, Andrea. *The Invention of Nature: Alexander von Humboldt's New World*. New York: Knopf Doubleday, 2015.

Yeo, Richard R. "Scientific Method and the Rhetoric of Science in Britain, 1830–1917." In *The Politics and Rhetoric of Scientific Method: Historical Studies*, edited by John A. Schuster and Richard R. Yeo, 259–97. Dordrecht: Springer, 1986.

INDEX

103, 128, 152; Clerkenwell, 106; Covent
Garden, 142; Crystal Palace, 151; Hanover
Square, 50; Holborn, 72, 104; Notting Hill,
74; Russell Square, 49; Regents Park, 40;
Tylor's visit to London, 51–53, 53, 63, 67,
78–82, 134, 166; Victoria, 58, 144
London Institution, 121
London Joint Stock Bank, 128
London Mechanics Institute, 19
Lough Erne, 111
Lubbock, John, 26–27
Lynes, Fanny, 104–7. *See also* Cock Lane
ghost; Kent, Elizabeth; Kent, William;
Scratching Fanny
Lyons, Jane, 143. *See also* Home, Daniel
Dunglas

Macaulay, George, 107–8
magic, 84, 94, 112–16, 132, 163. *See also*
conjuror; magician
magician, 132, 148, 151, 159–61, 163. *See also*
conjuror; magic
A Magician among the Spirits, 161. *See also*
Houdini, Harry
magnetism, 70. *See also* mesmerism
Malay Archipelago, 22–23
Malinowski, Bronisław, 157, 161. *See also*
functionalism
Mann, Horatio, Earl of Orford, 140
Man's Place in Nature, 129–30
Maori *pah*, 90
Marett, Robert Ranulph, 52
Marxist historiography, 8
Maskelyne, John Nevil, 132, 151. *See also*
conjurer; magic; magician
materializations, 37, 49–50, 73, 75. *See also*
Cook, Florence
Mather, Cotton, 115. *See also* witchcraft trials
McCabe, Joseph, 126
McLennan, John Ferguson, 88
Meditations among the Tombs, 127. *See also*
Hervey, James
Medium and Daybreak, 66, 70, 73–74. *See also*
Harrison, William Henry
Merton College, University of Oxford, 87
Mesmer, Franz, 72. *See also* magnetism;
mesmerism
mesmerism, 21–22, 70, 72, 75, 79, 121. *See also*

magnetism; Mesmer, Franz
Methodists, 140
Mexico, 54–55
microhistory, 11–12, 86, 104
middling-sort class, 53, 55, 87
miracles, 9, 29, 58, 90
Miracles and Modern Spiritualism, 13–14, 16,
18, 21, 46–48, 51, 62; credible witnessing, 19,
31–39; fact-based knowledge, 27–31; spirit
photography, 43–46; theory of spiritualism,
24–27. *See also* Wallace, Alfred Russel
Mirville, Charles-Jules de, 113, 115. *See also*
Cideville poltergeist
misperception, 47, 57, 134, 139, 141
Modern Spiritualism, 150. *See also* Podmore,
Frank
monogenesis, 18, 27, 54–56. *See also* polygenesis
More, Henry, 91
Morgan, Annie Owen, 49
Morgan, Henry Owen 49
Morris, Miss, 102–3
Moses, William Stainton, 78–80, 82, 115,
145–46
Müller, Friedrich Max, 88
multiple contextualizations, 11
Mumler, William Howard, 154–56. *See also*
spirit photography
Murray, Gilbert, 139
Museum of Practical Geology, 129
Myers, Frederic W. H., 97, 100, 102, 122, 134,
136
Les Mystéres de la Science, 118–19. *See also*
Figuier, Louis
The Mystery Revealed, 104. *See also* Goldsmith,
Oliver
Myth, Ritual, and Religion, 83, 88. *See also*
Lang, Andrew
mythology, 87–88, 92, 95

natural laws, 89
natural selection, 16, 26, 166; survival of
the fittest, 26. *See also* Darwin, Charles;
Darwinism; *Origin of Species*
The New Revelation, 162. *See also* Arthur
Conan Doyle
New York City, 59, 152, 154, 156
New York State, 22, 34, 110, 139
Newgate prison, 108, 141